THE COMPLETE IB BIOLOGY EXAM PREPARATION FOR SL & HL

Dr. Raquel Flemming Goulds

11 February 2017

ISBN-13:
978-1539598152

ISBN-10:
1539598152

INGENIUM PRESS, LTD.

Copyright 2017

All Rights Reserved Worldwide

Dedicated to all who do scientific research:

"If we knew what it was we were doing, it would not be called research, would it?"
— *Albert Einstein*

TABLE OF CONTENTS

- IB Biology . . . Waazzat? 11
- The IB Biology Curriculum Outlined 12
- The Essence of the Subject Matter 13
- Basic Concepts in Biology 14
- Homeostasis 14
- Unity 14
- Evolution 15
- Diversity 16
- Taxonomy 17
- Behavior and Interrelationships 17
- Continuity 17
- The Study of Structure 17
- Cells and Their Constituents 18
- Tissues and Organs 18
- The History of Biology 19
- The Greco-Roman World 20
- Aristotelian Concepts 20
- Botanical Investigations 21
- Post-Grecian Biological Studies 23
- Development of Botany and Zoology 23
- The Renaissance 24
- Resurgence of Biology 24

Advances in Botany 25
Advances in Anatomy 26
Advances in the 20th Century 27
The Establishment of Scientific Societies..................................... 28
Malpighi's Animal and Plant Studies 28
The Discovery of Cells 29
The Use of Structure for Classifying Organisms........................... 30
Reorganization of Groups of Organisms....................................... 30
The Development of Comparative Biological Studies 31
The Origins of Primordial Life 32
Biological Expeditions 33
The Development of Cell Theory 34
The Theory of Evolution... 35
The Study of the Reproduction and Development of Organisms . 36
Preformation versus Epigenesis.. 36
The Fertilization Process 36
The Study of Heredity 37
Pre-Mendelian Theories of Heredity 37
Mendelian Laws of Heredity 38
Elucidation of the Hereditary Mechanism.................................... 38
Biology in the 20th and 21st Centuries 39
Important Conceptual and Technological Developments............. 39
Intradisciplinary and Interdisciplinary Work 41

Changing Social and Scientific Values 43

Coping with the Problems of the Future 43

On to the Nitty-Gritty 44

IB Biology SL and HL Core 44

Topic #2: Molecular Biology - 21 Hours for Both SL and HL 46

Topic #3: Genetics - 15 Hours for Both SL and HL 51

Topic #4: Ecology - 12 Hours for Both SL and HL 55

Topic #5: Evolution and Biodiversity - 12 Hours for Both SL and HL 60

Topic #6: Human Physiology - 20 Hours for Both SL and HL 63

Additional Higher Level Topics 70

Topic #7: Nucleic Acids - 9 Hours for HL Only 70

Topic #8: Metabolism, Cell Respiration, and Photosynthesis - 14 Hours for HL Only 72

Topic #9: Plant Biology - 13 Hours for HL Only 75

Topic #10: Genetics and Evolution - 8 Hours for HL Only 77

Topic #11: Animal Physiology - 16 Hours for HL Only 78

Options 83

Option A: Neurobiology and Behavior - 15 Hours for SL and HL .. 83

Additional Higher Level Topics - Additional 10 Hours for HL 86

Option B: Biotechnology and Bioinformatics - 15 Hours for SL and HL 88

Additional Higher Level Topics - Additional 10 Hours for HL 91

Option C: Ecology and Conservation - 15 Hours for SL and HL ... 92

Additional Higher Level Topics - Additional 10 Hours for HL 95

Option D: Human Physiology - 15 Hours for SL and HL 96

Additional Higher Level Topics - Additional 10 Hours for HL 100

Practical Scheme of Work 101

Self-Test 102

Know these Definitions 102

Sources 125

IB Biology . . . Waazzat?

Biologists investigate the living world at all levels using many different approaches and techniques.

At one end of the scale is the cell, its molecular construction and complex metabolic reactions. At the other end of the scale biologists investigate the interactions that make whole ecosystems function. Many discoveries remain to be made and great progress is expected in the 21st century.

Through studying a science subject students should become aware of how scientists work and communicate with each other. While the scientific method may take on a wide variety of forms, the emphasis on a practical approach. In addition, through the overarching theme of the "Nature of Science" this knowledge and skills will be put into the context of way science and scientists work in the 21st Century and the ethical debates and limitations of creative scientific endeavour.

The sciences are taught practically. Students have opportunities to design investigations, collect data, develop manipulative skills, analyze results, collaborate with peers and evaluate and communicate their findings. The investigations may be laboratory based or they may make use of simulations and data bases. Students develop the skills to work independently on their own design, but also collegiately, including collaboration with schools in different regions, to mirror the way in which scientific research is conducted in the wider community.

I believe that thirty million of these animalcules together would not take up as much room, or be as big, as a coarse grain of sand.

~ ANTONIE VAN LEEUWENHOEK, 1632 TO 1723

Biologist

The IB Biology Curriculum Outlined

Higher level (240 hours)

- Internal assessment (individual investigation): 20%

- External assessment: 80%

Standard level (150 hours)

- Internal assessment (individual investigation): 20%

- External assessment: 80%

Key features of the curriculum and assessment models

- Available at standard (SL) and higher levels (HL)

- The minimum prescribed number of hours is 150 for SL and 240 for HL

- Students are assessed both externally and internally

- Biology students at SL and HL undertake a common core syllabus and a common internal assessment (IA) scheme.

- While there are core skills and activities common to both SL and HL students, students at HL are required to study the options and some topics in greater depth as well as some additional topics. The distinction between SL and HL is one of breadth and depth.

- A practical approach to the course delivery is emphasized through the interdisciplinary group 4 project and a mixture of both short-term and long-term experiments and investigations.

- Internal assessment accounts for 20% of the final assessment and this is assessed through a single individual investigation. This investigation may involve a hands-on approach, use of data-bases, modeling, simulation or a hybrid. Student work is internally assessed by the teacher and externally moderated by the IB.

The external assessment of biology consists of three written papers. In paper 1 there are 30 (at SL) or 40 (at HL) multiple-choice questions. Paper 2 contains short-answer and extended-response questions on the core (and Additional Higher Level (AHL) material at HL). Paper 3 has two sections; Section A contains one data-based question and several short-answer questions on experimental work on the core (and AHL material at HL). Section B contains short-answer and extended-response questions from each of the four options.[1]

The Essence of the Subject Matter

Biology, study of living things and their vital processes. The field deals with all the physicochemical aspects of life. The modern tendency toward cross-disciplinary research and the unification of scientific knowledge and investigation from different fields has resulted in significant overlap of the field of biology with other scientific disciplines. Modern principles of other fields—chemistry, medicine, and physics, for example—are integrated with those of biology in areas such as biochemistry, biomedicine, and biophysics.

Biology is subdivided into separate branches for convenience of study, though all the subdivisions are interrelated by basic principles. Thus, while it is custom to separate the study of plants (botany) from that of animals (zoology), and the study of the structure of organisms (morphology) from that of function (physiology), all living things share in common certain biological phenomena—for example, various means of reproduction, cell division, and the transmission of genetic material.

Biology is often approached on the basis of levels that deal with fundamental units of life. At the level of molecular biology, for example, life is regarded as a manifestation of chemical and energy transformations that occur among the many chemical constituents that compose an organism. As a result of the development of increasingly powerful and precise laboratory instruments and techniques, it is possible to understand and define with high precision and accuracy not only the ultimate physiochemical organization (ultrastructure) of the molecules in living matter but also the way living matter reproduces at the molecular level. Especially crucial to those advances was the rise of genomics in the late 20th and early 21st centuries.

---

[1] Much of this information is taken directly from the biology subject guide, available to all IB teachers on the Online Curriculum Centre (OCC).

Cell biology is the study of cells—the fundamental units of structure and function in living organisms. Cells were first observed in the 17th century, when the compound microscope was invented. Before that time, the individual organism was studied as a whole in a field known as organismic biology; that area of research remains an important component of the biological sciences. Population biology deals with groups or populations of organisms that inhabit a given area or region. Included at that level are studies of the roles that specific kinds of plants and animals play in the complex and self-perpetuating interrelationships that exist between the living and the nonliving world, as well as studies of the built-in controls that maintain those relationships naturally. Those broadly based levels—molecules, cells, whole organisms, and populations—may be further subdivided for study, giving rise to specializations such as morphology, taxonomy, biophysics, biochemistry, genetics, epigenetics, and ecology. A field of biology may be especially concerned with the investigation of one kind of living thing—for example, the study of birds in ornithology, the study of fishes in ichthyology, or the study of microorganisms in microbiology.

Basic Concepts in Biology

Homeostasis

The concept of homeostasis—that living things maintain a constant internal environment—was first suggested in the 19th century by French physiologist Claude Bernard, who stated that "all the vital mechanisms, varied as they are, have only one object: that of preserving constant the conditions of life."

As originally conceived by Bernard, homeostasis applied to the struggle of a single organism to survive. The concept was later extended to include any biological system from the cell to the entire biosphere, all the areas of Earth inhabited by living things.

Unity

All living organisms, regardless of their uniqueness, have certain biological, chemical, and physical characteristics in common. All, for example, are composed of basic units known as cells and of the same chemical substances, which, when analyzed, exhibit noteworthy similarities, even in such disparate organisms as bacteria and humans. Furthermore, since the action of any organism is determined by the manner in which its cells interact and since all cells interact in much the same way, the basic functioning of all organisms is also similar.

There is not only unity of basic living substance and functioning but also unity of origin of all living things. According to a theory proposed in 1855 by German pathologist Rudolf Virchow, "all living cells arise from pre-existing living cells." That theory appears to be true for all living things at the present time under existing environmental conditions. If, however, life originated on Earth more than once in the past, the fact that all organisms have a sameness of basic structure, composition, and function would seem to indicate that only one original type succeeded.

A common origin of life would explain why in humans or bacteria—and in all forms of life in between—the same chemical substance, deoxyribonucleic acid (DNA), in the form of genes accounts for the ability of all living matter to replicate itself exactly and to transmit genetic information from parent to offspring. Furthermore, the mechanisms for that transmittal follow a pattern that is the same in all organisms.

Whenever a change in a gene (a mutation) occurs, there is a change of some kind in the organism that contains the gene. It is this universal phenomenon that gives rise to the differences (variations) in populations of organisms from which nature selects for survival those that are best able to cope with changing conditions in the environment.

Evolution

In his theory of natural selection, which is discussed in greater detail later, Charles Darwin suggested that "survival of the fittest" was the basis for organic evolution (the change of living things with time). Evolution itself is a biological phenomenon common to all living things, even though it has led to their differences. Evidence to support the theory of evolution has come primarily from the fossil record, from comparative studies of structure and function, from studies of embryological development, and from studies of DNA and RNA (ribonucleic acid).

Diversity

Despite the basic biological, chemical, and physical similarities found in all living things, a diversity of life exists not only among and between species but also within every natural population. The phenomenon of diversity has had a long history of study because so many of the variations that exist in nature are visible to the eye. The fact that organisms changed during prehistoric times and that new variations are constantly evolving can be verified by paleontological records as well as by breeding experiments in the laboratory. Long after Darwin assumed that variations existed, biologists discovered that they are caused by a change in the genetic material (DNA). That change can be a slight alteration in the sequence of the constituents of DNA (nucleotides), a larger change such as a structural alteration of a chromosome, or a complete change in the number of chromosomes. In any case, a change in the genetic material in the reproductive cells manifests itself as some kind of structural or chemical change in the offspring. The consequence of such a mutation depends upon the interaction of the mutant offspring with its environment.

It has been suggested that sexual reproduction became the dominant type of reproduction among organisms because of its inherent advantage of variability, which is the mechanism that enables a species to adjust to changing conditions. New variations are potentially present in genetic differences, but how preponderant a variation becomes in a gene pool depends upon the number of offspring the mutants or variants produce (differential reproduction). It is possible for a genetic novelty (new variation) to spread in time to all members of a population, especially if the novelty enhances the population's chances for survival in the environment in which it exists. Thus, when a species is introduced into a new habitat, it either adapts to the change by natural selection or by some other evolutionary mechanism or eventually dies off. Because each new habitat means new adaptations, habitat changes have been responsible for the millions of different kinds of species and for the heterogeneity within each species.

The total number of extant animal and plant species is estimated at between roughly 5 million and 10 million; about 1.5 million of those species have been described by scientists. The use of classification as a means of producing some kind of order out of the staggering number of different types of organisms appeared as early as the book of Genesis—with references to cattle, beasts, fowl, creeping things, trees, and so on. The first scientific attempt at classification, however, is attributed to the Greek philosopher Aristotle, who tried to establish a system that would indicate the relationship of all things to each other. He arranged everything along a scale, or "ladder of nature," with nonliving things at the bottom; plants were placed below animals, and humankind was at the top. Other schemes that have been used for grouping species include large anatomical similarities, such as wings or fins, which indicate a natural relationship, and also similarities in reproductive structures.

Taxonomy

Taxonomy has been based on two major assumptions: one is that similar body construction can be used as a criterion for a classification grouping; the other is that, in addition to structural similarities, evolutionary and molecular relationships between organisms can be used as a means for determining classification.

Behavior and Interrelationships

The study of the relationships of living things to each other and to their environment is known as ecology. Because these interrelationships are so important to the welfare of Earth and because they can be seriously disrupted by human activities, ecology has become an important branch of biology.

Continuity

Whether an organism is a human or a bacterium, its ability to reproduce is one of the most important characteristics of life. Because life comes only from preexisting life, it is only through reproduction that successive generations can carry on the properties of a species.

The Study of Structure

Living things are defined in terms of the activities or functions that are missing in nonliving things. The life processes of every organism are carried out by specific materials assembled in definite structures. Thus, a living thing can be defined as a system, or structure, that reproduces, changes with its environment over a period of time, and maintains its individuality by constant and continuous metabolism.

Cells and Their Constituents

Biologists once depended on the light microscope to study the morphology of cells found in higher plants and animals. The functioning of cells in unicellular and in multicellular organisms was then postulated from observation of the structure; the discovery of the chloroplastids in the cell, for example, led to the investigation of the process of photosynthesis. With the invention of the electron microscope, the fine organization of the plastids could be used for further quantitative studies of the different parts of that process.

Qualitative and quantitative analyses in biology make use of a variety of techniques and approaches to identify and estimate levels of nucleic acids, proteins, carbohydrates, and other chemical constituents of cells and tissues. Many such techniques make use of antibodies or probes that bind to specific molecules within cells and that are tagged with a chemical, commonly a fluorescent dye, a radioactive isotope, or a biological stain, thereby enabling or enhancing microscopic visualization or detection of the molecules of interest.

Chemical labels are powerful means by which biologists can identify, locate, or trace substances in living matter. Some examples of widely used assays that incorporate labels include the Gram stain, which is used for the identification and characterization of bacteria; fluorescence in situ hybridization, which is used for the detection of specific genetic sequences in chromosomes; and luciferase assays, which measure bioluminescence produced from luciferin-luciferase reactions, allowing for the quantification of a wide array of molecules.

Tissues and Organs

Early biologists viewed their work as a study of the organism. The organism, then considered the fundamental unit of life, is still the prime concern of some modern biologists, and understanding how organisms maintain their internal environment remains an important part of biological research. To better understand the physiology of organisms, researchers study the tissues and organs of which organisms are composed. Key to that work is the ability to maintain and grow cells in vitro ("in glass"), otherwise known as tissue culture.

Some of the first attempts at tissue culture were made in the late 19th century. In 1885, German zoologist Wilhelm Roux maintained tissue from a chick embryo in a salt solution. The first major breakthrough in tissue culture, however, came in

1907 with the growth of frog nerve cell processes by American zoologist Ross G. Harrison. Several years later, French researchers Alexis Carrel and Montrose Burrows had refined Harrison's methods and introduced the term *tissue culture*. Using stringent laboratory techniques, workers have been able to keep cells and tissues alive under culture conditions for long periods of time. Techniques for keeping organs alive in preparation for transplants stem from such experiments.

Advances in tissue culture have enabled countless discoveries in biology. For example, many experiments have been directed toward achieving a deeper understanding of biological differentiation, particularly of the factors that control differentiation. Crucial to those studies was the development in the late 20th century of tissue culture methods that allowed for the growth of mammalian embryonic stem cells—and ultimately human embryonic stem cells—on culture plates.

The History of Biology

There are moments in the history of all sciences when remarkable progress is made in relatively short periods of time. Such leaps in knowledge result in great part from two factors: one is the presence of a creative mind—a mind sufficiently perceptive and original to discard hitherto accepted ideas and formulate new hypotheses; the second is the technological ability to test the hypotheses by appropriate experiments. The most original and inquiring mind is severely limited without the proper tools to conduct an investigation; conversely, the most-sophisticated technological equipment cannot of itself yield insights into any scientific process.

An example of the relationship between those two factors was the discovery of the cell. For hundreds of years there had been speculation concerning the basic structure of both plants and animals. Not until optical instruments were sufficiently developed to reveal cells, however, was it possible to formulate a general hypothesis, the cell theory, that satisfactorily explained how plants and animals are organized. Similarly, the significance of Gregor Mendel's studies on the mode of inheritance in the garden pea remained neglected for many years until technological advances made possible the discovery of the chromosomes and the part they play in cell division and heredity. Moreover, as a result of the relatively recent development of extremely sophisticated instruments, such as the electron microscope, the ultracentrifuge, and automated DNA sequencing machines, biology has moved from being a largely descriptive science—one concerned with entire cells and organisms—to a discipline that increasingly emphasizes the subcellular and molecular aspects of organisms and attempts to equate structure with function at all levels of biological organization.

The Greco-Roman World

Although the Babylonians, Assyrians, Egyptians, Chinese, and Indians amassed much biological information, they lived in a world believed to be dominated by unpredictable demons and spirits. Hence, learned individuals in those early cultures directed their studies toward an understanding of the supernatural, rather than the natural, world. Anatomists, for example, dissected animals not to gain an understanding of their structure but to study their organs in order to predict the future. With the emergence of the Greek civilization, however, those mystical attitudes began to change. Around 600 bce there arose a school of Greek philosophers who believed that every event has a cause and that a particular cause produces a particular effect. That concept, known as causality, had a profound effect on subsequent scientific investigation. Furthermore, those philosophers assumed the existence of a "natural law" that governs the universe and can be comprehended by humans through the use of their powers of observation and deduction. Although they established the science of biology, the greatest contribution the Greeks made to science was the idea of rational thought.

Aristotelian Concepts

Around the middle of the 4th century bce, ancient Greek science reached a climax with Aristotle, who was interested in all branches of knowledge, including biology. Using his observations and theories, Aristotle was the first to attempt a system of animal classification, in which he contrasted animals containing blood with those that were bloodless. The animals with blood included those now grouped as mammals (except the whales, which he placed in a separate group), birds, amphibians, reptiles, and fishes. The bloodless animals were divided into the cephalopods, the higher crustaceans, the insects, and the testaceans, the last group being a collection of all the lower animals. His careful examination of animals led to the understanding that mammals have lungs, breathe air, are warm-blooded, and suckle their young. Aristotle was the first to show an understanding of an overall systematic taxonomy and to recognize units of different degrees within the system.

The most-important part of Aristotle's work was that devoted to reproduction and the related subjects of heredity and descent. He identified four means of reproduction, including the abiogenetic origin of life from nonliving mud, a belief held by Greeks of that time. Other modes of reproduction recognized by him included budding (asexual reproduction), sexual reproduction without copulation, and sexual reproduction with copulation. Aristotle described sperm and ova and believed that the menstrual blood of viviparous organisms (those that give birth to living young) was the actual generative substance.

Although Aristotle recognized that species are not stable and unalterable and although he attempted to classify the animals he observed, he was far from developing any pre-Darwinian ideas concerning evolution. In fact, he rejected any suggestion of natural selection and sought teleological explanations (i.e., all phenomena in nature are shaped by a purpose) for any given observation. Nevertheless, many important scientific principles, some of which are often thought of as 20th-century concepts, can be ascribed to Aristotle. The following are a few such: (1) Using birds as an example, he formulated the principle that all organisms are structurally and functionally adapted to their habits and habitats. (2) Nature is parsimonious; it does not expend unnecessary energy. (3) In classifying animals, Aristotle rejected the idea of dividing them solely by their external structures (e.g., animals with wings and those without wings). He recognized instead a basic unity of plan among diverse organisms, a principle that is still conceptually and scientifically sound. Further, Aristotle also believed that the entire living world could be described as a unified organization rather than as a collection of diverse groups. (4) By his observations, Aristotle realized the importance of structural homology, basically similar organs in different animals, and functional analogy, different structures that serve somewhat the same function—e.g., the hand, the claw, and the hoof are analogous structures. Those principles constitute the basis for the biological field of study known as comparative anatomy. (5) Aristotle's observations also led to the formulation of the principle that general structures appear before specialized ones and that tissues differentiate before organs.

Botanical Investigations

Of all the works of Aristotle that have survived, none deals with what was later differentiated as botany, although it is believed that he wrote at least two treatises on plants. Fortunately, however, the work of Theophrastus, one of Aristotle's students, has been preserved to represent plant science of the Greek period. Like Aristotle, Theophrastus was a keen observer, although his works do not express the depth of original thought exemplified by his teacher. In his great work, *De historia et causis plantarum* (*The Calendar of Flora*, 1761), in which the morphology, natural history, and therapeutic use of plants are described, Theophrastus distinguished between the external parts, which he called organs, and the internal parts, which he called tissues. This was an important achievement because Greek scientists of that period had no established scientific terminology for specific structures. For that reason, both Aristotle and Theophrastus were obliged to write very long descriptions of structures that can be described rapidly and simply today. Because of that difficulty, Theophrastus sought to develop a scientific nomenclature by giving special meaning to words that were then in more or less current use; for example, *karpos* for fruit and *perikarpion* for seed vessel.

Although he did not propose an overall classification system for plants, more than 500 of which are mentioned in his writings, Theophrastus did unite many species into what are now considered genera. In addition to writing the earliest detailed description of how to pollinate the date palm by hand and the first unambiguous account of sexual reproduction in flowering plants, he also recorded observations on seed germination and development.

If physics and biology one day meet, and one of the two is swallowed up, that one will be biology.

~ *J. B. S. HALDANE, 1892 TO 1964*

Biologist

Post-Grecian Biological Studies

With Aristotle and Theophrastus, the great Greek period of scientific investigation came to an end. The most famous of the new centres of learning were the library and museum in Alexandria. From 300 bce until around the time of Christ, all significant biological advances were made by physicians at Alexandria. One of the most outstanding of those individuals was Herophilus, who dissected human bodies and compared their structures with those of other large mammals. He recognized the brain, which he described in detail, as the centre of the nervous system and the seat of intelligence. On the basis of his knowledge, he wrote a general anatomical treatise, a special one on the eyes, and a handbook for midwives.

Erasistratus, a younger contemporary and reputed rival of Herophilus who also worked at the museum in Alexandria, studied the valves of the heart and the circulation of blood. Although he was wrong in supposing that blood flows from the veins into the arteries, he was correct in assuming that small interconnecting vessels exist. He thus suspected (but did not see) the presence of capillaries; he thought, however, that the blood changed into air, or *pneuma*, when it reached the arteries, to be pumped throughout the body.

Perhaps the last of the ancient biological scientists of note was Galen of Pergamum, a Greek physician who practiced in Rome during the middle of the 2nd century ce. His early years were spent as a surgeon at the gladiatorial arena, which gave him the opportunity to observe details of human anatomy. At that time in Rome, however, it was considered improper to dissect human bodies, and, as a result, a detailed study of human anatomy was not possible. Thus, though Galen's research on animals was thorough, his knowledge of human anatomy was faulty. Because his work was extensive and clearly written, Galen's writings, nevertheless, dominated medicine for centuries.

Development of Botany and Zoology

During the 12th century the growth of biology was sporadic. Nevertheless, it was during that time that botany was developed from the study of plants with healing properties; similarly, from veterinary medicine and the pleasures of the hunt came zoology. Because of the interest in medicinal plants, herbs in general began to be described and illustrated in a realistic manner. Although Arabic science was well developed during the period and was far in advance of Latin, Byzantine, and

Chinese cultures, it began to show signs of decline. Latin learning, on the other hand, rapidly increasing, was best exemplified perhaps by the mid-13th-century German scholar Albertus Magnus (Saint Albert the Great), who was probably the greatest naturalist of the Middle Ages. His biological writings (*De vegetabilibus*, seven books, and *De animalibus*, 26 books) were based on the classical Greek authorities, predominantly Aristotle. But in spite of that classical basis, a significant amount of his work contained new observations and facts; for example, he described with great accuracy the leaf anatomy and venation of the plants he studied.

Albertus was particularly interested in plant propagation and reproduction and discussed in some detail the sexuality of plants and animals. Like his Greek predecessors, he believed in spontaneous generation; he also believed that animals were more perfect than plants, because they required two individuals for the sexual act. Perhaps one of Albertus's greatest contributions to medieval biology was the denial of many superstitions believed by his contemporaries, a skepticism that, together with the reintroduction of Aristotelian biology, was to have profound effects on subsequent European science.

One of Albertus's pupils was Thomas Aquinas, who, like his mentor, endeavoured to reconcile Aristotelian philosophy and the teachings of the church. Because Aquinas was a rationalist, he declared that God created the reasoning mind; hence, by true intellectual processes of reasoning, man could not arrive at a conclusion that was in opposition to Christian thought. Acceptance of this philosophy made possible a revival of rational learning that was consistent with Christian belief.

The Renaissance

Resurgence of Biology

Beginning in Italy during the 14th century, there was a general ferment within the culture itself, which, together with the rebirth of learning (partly as a result of the rediscovery of Greek work), is referred to as the Renaissance. Interestingly, it was the artists, rather than the professional anatomists, who were intent upon a true rendering of the bodies of animals, including humans, and thus were motivated to gain their knowledge firsthand by dissection. No individual better exemplifies the Renaissance than Leonardo da Vinci, whose anatomical studies of the human form during the late 1400s and early 1500s were so far in advance of the age that they included details not recognized until a century later. Furthermore, while dissecting animals and examining their structure, Leonardo compared them with the structure of humans. In doing so he was the first to indicate the homology between the arrangements of bones and joints in the leg of the human and that of the horse, despite the superficial differences. Homology was to become an important concept in uniting outwardly diverse groups of animals into distinct units, a factor that is of great significance in the study of evolution.

Other factors had a profound effect upon the course of biology in the 1500s, particularly the introduction of printing around the middle of the century, the increasing availability of paper, and the perfected art of the wood engraver, all of which meant that illustrations as well as letters could be transferred to paper. In addition, after the Turks conquered Byzantium in 1453, many Greek scholars took refuge in the West; the scholars of the West thus had direct access to the scientific works of antiquity rather than indirect access through Arabic translations.

Advances in Botany

Over the period 1530–40, German theologian and botanist Otto Brunfels published the two volumes of his *Herbarum vivae eicones*, a book about plants, which, with its fresh and vigorous illustrations, contrasted sharply with earlier texts, whose authors had been content merely to copy from old manuscripts. In addition to books on the same subject, Hieronymus Bock (Latinized to Tragus) and Leonhard Fuchs also published about the mid-1500s descriptive well-illustrated texts about common wild flowers. The books published by the three men, who are often referred to as the German fathers of botany, may be considered the forerunners of modern botanical floras (treatises on or lists of the plants of an area or period).

Throughout the 16th century, interest in botanical study also existed in other countries, including the Netherlands, Switzerland, Italy, and France. During that time there was a great improvement in the classification of plants, which had been described in ancient herbals merely as trees, shrubs, or plants and, in later books, were either listed alphabetically or arranged in some arbitrary grouping. The necessity for a systematic method to designate the increasing number of plants being described became obvious. Accordingly, using a binomial system very similar to modern biological nomenclature, the Swiss botanist Gaspard Bauhin designated plants by a generic and a specific name. Although affinities between plants were indicated by the use of common generic names, Bauhin did not speculate on their common kinship.

Pierre Belon, a French naturalist who traveled extensively in the Middle East, where he studied the flora, illustrates the wide interest of the 16th-century biologists. Although his botanical work was limited to two volumes, one on trees and one on horticulture, his books on travel included numerous biological entries. His two books on fishes reveal much about the state of systematics at the time, including that of not only fishes but also other aquatic creatures such as mammals, crustaceans, mollusks, and worms. In his *L'Histoire de la nature des oyseaux* (1555; "Natural History of Birds"), however, in which Belon's taxonomy was remarkably similar to that used in the modern era, he showed a clear grasp of comparative anatomy, particularly of the skeleton, publishing the first picture of a bird skeleton beside a human skeleton to point out the homologies. Numerous other European naturalists who traveled extensively also brought back accounts of exotic animals and plants, and most of them wrote voluminous records of their excursions. Two other factors contributed significantly to the development of botany at the time: first was the establishment of botanical gardens by the universities, as distinct from the earlier gardens that had been established for medicinal plants; second was the collection of dried botanical specimens, or herbaria.

It is perhaps surprising that the great developments in botany during the 16th century had no parallel in zoology. Instead, there arose a group of biologists known as the Encyclopedists, best represented by Conrad Gesner, a 16th-century Swiss naturalist, who compiled books on animals that were illustrated by some of the finest artists of the day (Albrecht Dürer, for example). But because the descriptions of many of the animals were grossly inaccurate, in many cases continuing the legends of the Greeks, apart from their aesthetic value the books did little to advance zoological knowledge.

Advances in Anatomy

Like that of botany, the beginning of the modern scientific study of anatomy can be traced to a combination of humanistic learning, Renaissance art, and the craft of printing. Although Leonardo da Vinci initiated anatomical studies of human cadavers, his work was not known to his contemporaries. Rather, the appellation father of modern human anatomy generally is accorded to the Belgian anatomist Andreas Vesalius, who studied initially at the rather conservative schools in Leuven (Louvain) and Paris, where he became a successful teacher very familiar with Galen's work. In 1537 he went to Padua, where he became noted for far-reaching teaching reforms. Most important, Vesalius abolished the practice of having someone else do the actual dissection; instead, he dissected his own cadavers and lectured to students from his findings. His text, *De humani corporis fabrica libri septem* (1543; "The Seven Books on the Structure of the Human Body"), was the most extensive and accurate work on the subject of anatomy at the time and, as such, constituted a foundation of great importance for biology. Perhaps Vesalius's greatest contribution, however, was that he inspired a group of younger scientists to be critical and to accept a description only after they had verified it. Thus, as anatomists became more questioning and critical of the works of others, the errors of Galen were exposed. Of Vesalius's successors, Michael Servetus, a Spanish theologian and physician, discovered the pulmonary circulation of the blood from the right chamber of the heart to the lungs and stated that the blood did not pass through the central septum (wall) of the heart, as had previously been believed.

Advances in the 20th Century

Seventeenth-century advances in biology included the establishment of scientific societies for the dissemination of ideas and progress in the development of the microscope, through which scientists discovered a hitherto invisible world that had far-reaching effects on biology. Systematizing and classifying, however, dominated biology throughout much of the 17th and 18th centuries, and it was during that time that the importance of the comparative study of living organisms, including humans, was realized. During the 18th century the long-held idea that living organisms could originate from nonliving matter (spontaneous generation) began to crumble, but it was not until after the mid-19th century that it was finally disproved by the French chemist and microbiologist Louis Pasteur, who demonstrated the self-replicating ability of microorganisms.

Biological expeditions added to the growing body of knowledge of plant and animal forms and led to the 19th-century development of the theory of evolution. The 19th century was one of great progress in biology: in addition to the formulation of the theory of evolution, the cell theory was established, the foundations for modern embryology were laid, and the laws of heredity were discovered.

The Establishment of Scientific Societies

A development of great importance to science was the establishment in Europe of academies or societies; they consisted of small groups of men who met to discuss subjects of mutual interest. Although some of the groups enjoyed the financial patronage of princes and other wealthy members of society, the members' interest in science was the sole sustaining force. The academies also provided freedom of expression, which, together with the stimulus of exchanging ideas, contributed greatly to the development of scientific thought. One of the earliest of these organizations was the Italian Accademia dei Lincei (Academy of the Lynx-eyed), founded in Rome around 1603. Galileo Galilei made a microscope for the society; another of its members, Johannes Faber, an entomologist, gave the instrument its name. Other academies in Europe included the French Academy of Sciences (founded in 1666), a German Academy in Leipzig, and a number of small academies in England that in 1662 became incorporated under royal charter as the Royal Society of London, an organization that was to have considerable influence on scientific developments in England.

In addition to providing a forum for the discussion of scientific matters, another important aspect of those societies was their publications. Before the advent of printing there were no convenient means for the wide dissemination of scientific knowledge and ideas; hence, scientists were not well informed about the works of others. To correct that deficiency in communications, the early academies initiated several publications, the first of which, *Journal des Savants* (originally *Journal des Sçavans*), was published in 1665 in France. Three months later, the Royal Society of London originated its *Philosophical Transactions*. At first the publication was devoted to reviews of work completed and in progress; later, however, the emphasis gradually changed to accounts of original investigations that maintained a high level of scientific quality. Gradually, specialized journals of science made their appearance, though not until at least another century had passed.

Malpighi's Animal and Plant Studies

The Italian biologist and physician Marcello Malpighi conducted extensive studies in animal anatomy and histology (the microscopic study of the structure, composition, and function of tissues). He was the first to describe the inner (malpighian) layer of the skin, the papillae of the tongue, the outer part (cortex) of the cerebral area of the brain, and the red blood cells. He wrote a detailed monograph on the silkworm; a further major contribution was a description of the development of the chick, beginning with the 24-hour stage. In addition to those and other animal studies, Malpighi made detailed investigations in plant anatomy. He systematically described the various parts of plants, such as bark, stem, roots, and seeds, and discussed processes such as germination and gall formation. Many of Malpighi's drawings of plant anatomy remained unintelligible to botanists until the structures were rediscovered in the 19th century. Although Malpighi was not a technical innovator, he does exemplify the functioning of the educated 17th-century mind, which, together with curiosity and patience, resulted in many advances in biology.

The Discovery of Cells

Of the five microscopists, Robert Hooke was perhaps the most intellectually preeminent. As curator of instruments at the Royal Society of London, he was in touch with all new scientific developments and exhibited interest in such disparate subjects as flying and the construction of clocks. In 1665 Hooke published his *Micrographia*, which was primarily a review of a series of observations that he had made while following the development and improvement of the microscope. Hooke described in detail the structure of feathers, the stinger of a bee, the radula, or "tongue," of mollusks, and the foot of the fly. It is Hooke who coined the word *cell*; in a drawing of the microscopic structure of cork, he showed walls surrounding empty spaces and referred to the structures as cells. He described similar structures in the tissue of other trees and plants and discerned that in some tissues the cells were filled with a liquid while in others they were empty. He therefore supposed that the function of the cells was to transport substances through the plant.

Although the work of any of the classical microscopists seems to lack a definite objective, it should be remembered that these men embodied the concept that observation and experiment were of prime importance, that mere hypothetical, philosophical speculations were not sufficient. It is remarkable that so few men, working as individuals totally isolated from each other, should have recorded so many observations of such fundamental importance. The great significance of their work was that it revealed, for the first time, a world in which living organisms display an almost incredible complexity.

Work with the compound microscope languished for nearly 200 years, mainly because the early lenses tended to break up white light into its constituent parts. That technical problem was not solved until the invention of achromatic lenses, which were introduced about 1830. In 1878 a modern achromatic compound microscope was produced from the design of the German physicist Ernst Abbe. Abbe subsequently designed a substage illumination system, which, together with the introduction of a new substage condenser, paved the way for the biological discoveries of that era.

The Use of Structure for Classifying Organisms

Two systematists of the 17th and 18th centuries were the English naturalist John Ray and the Swedish naturalist and explorer Carolus Linnaeus. Ray, who studied at Cambridge, was particularly interested in the work of the ancient compilers of herbals, especially those who had attempted to formulate some means of classification. Recognizing the need for a classification system that would apply to both plants and animals, Ray employed in his classification schemes extremely precise descriptions for genera and species. By basing his system on structures, such as the arrangement of toes and teeth in animals, rather than colour or habitat, Ray introduced a new and very important concept to taxonomic biology.

Reorganization of Groups of Organisms

Prior to Linnaeus, most taxonomists started their classification systems by dividing all the known organisms into large groups and then subdividing them into progressively smaller groups. Unlike his predecessors, Linnaeus began with the species, organizing them into larger groups or genera, and then arranging analogous genera to form families and related families to form orders and classes. Probably utilizing the earlier work of Grew and others, Linnaeus chose the structure of the reproductive organs of the flower as a basis for grouping the higher plants. Thus, he distinguished between plants with real flowers and seeds (phanerogams) and those lacking real flowers and seeds (cryptogams), subdividing the former into hermaphroditic (bisexual) and unisexual forms. For animals, following Ray's work, Linnaeus relied upon teeth and toes as the basic characteristics of mammals; he used the shape of the beak as the basis for bird classification. Having demonstrated that a binomial classification system based on concise and accurate descriptions could be used for the grouping of organisms, Linnaeus established taxonomic biology as a discipline.

Later developments in classification were initiated by the French biologists Comte de Buffon, Jean-Baptiste Lamarck, and Georges Cuvier, all of whom made lasting contributions to biological science, particularly in comparative studies. Subsequent systematists have been chiefly interested in the relationships between animals and have endeavored to explain not only their similarities but also their differences in broad terms that encompass, in addition to structure, composition, function, genetics, evolution, and ecology.

The Development of Comparative Biological Studies

Once the opprobrium attached to the dissection of human bodies had been dispelled in the 16th century, anatomists directed their efforts toward a better understanding of human structure. In doing so they generally ignored other animals, at least until the latter part of the 17th century, when biologists began to realize that important insights could be gained by comparative studies of all animals, including humans. One of the first of such anatomists was the English physician Edward Tyson, who studied the anatomy of an immature chimpanzee in detail and compared it with that of a human. In making further comparisons between the chimpanzee and other primates, Tyson clearly recognized points of similarity between those animals and humans. Not only was this a major contribution to physical anthropology, but it was also an indication—nearly two centuries before Darwin—of the existence of relationships between humans and other primates.

Among those who gave comparative studies their greatest impetus was Georges Cuvier, who utilized large collections of biological specimens sent to him from all over the world to work out a systematic organization of the animal kingdom. In addition to establishing a connection between systematic and comparative anatomy, he believed that there was a "correlation of parts" according to which a given type of structure (e.g., feathers) is related to a certain anatomical formation (e.g., a wing), which in turn is related to other specific formations (e.g., the clavicle), and so on. In other words, he felt that a great deal of anatomical information could be deduced about an organism even if the whole specimen was not available. That insight was to be of great practical importance in the study of fossils, in which Cuvier played a leading role. Indeed, the 1812 publication of Cuvier's *Recherches sur les ossemens fossiles de quadrupèdes* (translated as *Research on Fossil Bones* in 1835) laid the foundation for the science of paleontology. But in order to reconcile his scientific findings with his personal religious beliefs, Cuvier postulated a series of catastrophic events that could account for both the presence of fossils and the immutability of existing species.

Biology is the study of the complex things in the Universe. Physics is the study of the simple ones.

~ *RICHARD DAWKINS, 1941 – PRESENT*

Evolutionary Biologist

The Origins of Primordial Life

In the 1920s the Russian biochemist Aleksandr Oparin and other scientists suggested that life may have come from nonliving matter under conditions that existed on primitive Earth, when the atmosphere consisted of the gases methane, ammonia, water vapour, and hydrogen. According to that concept, energy supplied by electrical storms and ultraviolet light may have broken down the atmospheric gases into their constituent elements, and organic molecules may have been formed when the elements recombined.

Some of those ideas have been verified by advances in geochemistry and molecular genetics; experimental efforts have succeeded in producing amino acids and proteinoids (primitive protein compounds) from gases that may have been present on Earth at its inception, and amino acids have been detected in rocks that are more than three billion years old. With improved techniques it may be possible to produce precursors of or actual self-replicating living matter from nonliving substances. But whether it is possible to create the actual living heterotrophic forms from which autotrophs supposedly developed remains to be seen.

Biological Expeditions

Although a number of 16th- and 17th-century travelers provided much valuable information about the plants and animals in Asia, America, and Africa, most of that information was collected by curious individuals rather than trained observers. In the 18th and 19th centuries, however, such information was collected increasingly in the course of organized scientific expeditions, usually under the auspices of a particular government. The most notable of those efforts were the voyages of the ships known as the HMS *Endeavour*, the HMS *Investigator*, the HMS *Beagle*, and the HMS *Challenger*, all sponsored by the English government.

Capt. James Cook sailed the *Endeavour* to the South Pacific islands, New Zealand, New Guinea, and Australia in 1768; the voyage provided the British naturalist and explorer Joseph Banks with the opportunity to make a very extensive collection of plants and notes, which helped establish him as a leading biologist. Another expedition to the same area in the *Investigator* in 1801 included the Scottish botanist Robert Brown, whose work on the plants of Australia and New Zealand became a classic; especially important were his descriptions of how certain plants adapt to different environmental conditions. Brown is also credited with discovering the cell nucleus and analyzing sexual processes in higher plants.

One of the most-famous biological expeditions of all time was that of the *Beagle* (1831–36), on which Charles Darwin served as naturalist. Although Darwin's primary interest at the time was geology, his visit to the Galápagos Islands aroused his interest in biology and caused him to speculate about their curious insular animal life and the significance of isolation in space and time for the formation of species. During the *Beagle* voyage, Darwin collected specimens of and accumulated copious notes on the plants and animals of South America and Australia, for which he received great acclaim on his return to England.

The voyage of the *Challenger* (*see* Challenger Expedition) from 1872 to 1876 was organized by the British Admiralty to study oceanography, meteorology, and natural history. Under the leadership of the Scottish naturalist Charles Wyville Thomson, vast collections of plants and animals were made, the importance of plankton (minute free-floating aquatic organisms) as a source of food for larger marine organisms was recognized, and many new planktonic species were discovered. A particularly significant aspect of the *Challenger* voyage was the interest it stimulated in the new science of marine biology.

In spite of those expeditions, the contributions made by individuals were still very important. The British naturalist Alfred Russel Wallace, for example, undertook explorations of the Malay Archipelago from 1854 to 1862. In 1876 he published his book *The Geographical Distribution of Animals*, in which he divided the landmasses into six zoogeographical regions and described their characteristic fauna. Wallace also contributed to the theory of evolution, publishing in 1870 a book expressing his views, *Contributions to the Theory of Natural Selection*.

The Development of Cell Theory

Although the microscopists of the 17th century had made detailed descriptions of plant and animal structure and though Hooke had coined the term *cell* to describe the compartments he had observed in cork tissue, their observations lacked an underlying theoretical unity. It was not until 1838 that the German botanist Matthias Jacob Schleiden, interested in plant anatomy, stated that "the lower plants all consist of one cell, while the higher ones are composed of (many) individual cells." When the German physiologist Theodor Schwann, Schleiden's friend, extended the cellular theory to include animals, he thereby brought about a rapprochement between botany and zoology. The formation of the cell theory—all plants and animals are made up of cells—marked a great conceptual advance in biology, and it resulted in renewed attention to the living processes that go on in cells.

In 1846, after several investigators had described the streaming movement of the cytoplasm in plant cells, the German botanist Hugo von Mohl coined the word *protoplasm* to designate the living substance of the cell. The concept of protoplasm as the physical basis of life led to the development of cell physiology.

A further extension of the cell theory was the development of cellular pathology by the German scientist Rudolf Virchow, who established the relationship between abnormal events in the body and unusual cellular activities. Virchow's work gave a new direction to the study of pathology and resulted in advances in medicine.

The detailed description of cell division was contributed by the German plant cytologist Eduard Strasburger, who observed the mitotic process in plant cells and further demonstrated that nuclei arise only from preexisting nuclei. Parallel work in mammals was carried out by the German anatomist Walther Flemming, who published his most important findings in *Zellsubstanz, Kern und Zelltheilung* (""Cell Substance, Nucleus and Cell Division"") in 1882.

The Theory of Evolution

As knowledge of plant and animal forms accumulated during the 16th, 17th, and 18th centuries, a few biologists began to speculate about the ancestry of those organisms, though the prevailing view was that promulgated by Linnaeus—namely, the immutability of the species. Among the early speculations voiced during the 18th century, the British physician Erasmus Darwin (grandfather of Charles Darwin), concluded that species descend from common ancestors and that there is a struggle for existence among animals. The French biologist Jean-Baptiste Lamarck, among the most important of the 18th-century evolutionists, recognized the role of isolation in species formation; he also saw the unity in nature and conceived the idea of the evolutionary tree.

A complete theory of evolution was not announced, however, until the publication in 1859 of Charles Darwin's *On the Origin of Species by Means of Natural Selection or the Preservation of Favoured Races in the Struggle for Life*. In his book Darwin stated that all living creatures multiply so rapidly that if left unchecked they would soon overpopulate the world. According to Darwin, the checks on population size are maintained by competition for the means of life. Hence, if any member of a species differs in some way that makes it better fitted to survive, then it will have an advantage that its offspring would be likely to perpetuate. Darwin's work reflects the influence of the British economist Thomas Robert Malthus, who in 1838 published an essay on population in which he warned that if humans multiply more rapidly than their food supply, competition for existence will result. Darwin was also influenced by the British geologist Charles Lyell, who realized from his studies of geological formations that the relative ages of deposits could be estimated by means of the proportion of living and extinct mollusks. But it was not until after his travels aboard the *Beagle* (1831–36), during which he observed a great richness and diversity of island fauna, that Darwin began to develop his theory of evolution. Alfred Russel Wallace had reached conclusions similar to those of Darwin following his studies of plants and animals in the Malay Archipelago. A short paper dealing with this subject sent by Wallace to Darwin finally resulted in the publication of Darwin's own theories.

Conceptually, the theory was of the utmost significance, accounting as it did for the formation of new species. Following the subsequent discovery of the chromosomal basis of inheritance and the laws of heredity, it could be seen that natural selection does not involve the sharp alternatives of life or death but results from the differential survival of variants. Today the universal principle of natural selection, which is the central concept of Darwin's theory, is firmly established.

The Study of the Reproduction and Development of Organisms

Preformation versus Epigenesis

A question posed by Aristotle was whether the embryo is preformed and therefore only enlarges during development or whether it differentiates from an amorphous beginning. Two conflicting schools of thought had been based on that question: the preformation school maintained that the egg contains a miniature individual that develops into the adult stage in the proper environment; the epigenesis school believed that the egg is initially undifferentiated and that development occurs as a series of steps. Prominent supporters of the preformation doctrine, which was widely held until the 18th century, included Malpighi, Swammerdam, and Leeuwenhoek. In the 19th century, as criticism of preformation mounted, the Prussian Estonian embryologist Karl Ernst von Baer provided the final evidence against the theory. His discovery of the mammalian egg and his recognition of the formation of the germ layers out of which the embryonic organs develop laid the foundations of modern embryology.

The Fertilization Process

Despite the many early descriptions of spermatozoa, their essential role in fertilization was not proved until 1879, when the Swiss physician and zoologist Hermann Fol observed the penetration of a spermatozoon into an ovum. Prior to that discovery, during the period from 1823 to 1830, the existence of the sexual process in flowering plants had been demonstrated by the Italian astronomer and optician Giovanni Battista Amici and confirmed by others. The discovery of fertilization in plants was of great importance to the development of plant hybrids, which are produced by cross-pollination between different species; it was also of great significance to the studies of genetics and evolution.

The universal occurrence and remarkable similarity of the fertilization process, regardless of the organism in which it occurs, provoked many of the leading investigators of the time to search for the underlying mechanism. It was realized that there must be some way by which the number of chromosomes is reduced before fertilization; otherwise, the chromosome number would double every time a sperm fused with an egg. In 1883 the Belgian embryologist and cytologist Edouard van Beneden showed that the eggs and the sperm in the worm *Ascaris* contain half the number of chromosomes found in the body cells. To account for the halving of the chromosomes in the sex cells, a process known as meiosis, in 1887 the German biologist August Weismann suggested that there must be two different types of cell division, and by 1900 the details of meiosis had been elucidated.

The Study of Heredity

Pre-Mendelian Theories of Heredity

The fundamental laws of heredity were discovered in 1865 by the Austrian botanist, teacher, and Augustinian prelate Gregor Mendel, though his work was ignored until its rediscovery in 1900. There were, however, a number of views on the subject that had been expressed long before Mendel. The Greek philosophers, for example, believed that the traits of individuals were acquired from contact with the environment and that such acquired characteristics could be inherited by offspring. Because Lamarck was the most famous proponent of the inheritance of acquired characteristics, the theory is called Lamarckism. This concept, which emphasized the use and disuse of organs as the significant factor in determining the characteristics of an individual, postulated that any alterations in the individual could be transmitted to the offspring through the gametes.

In 1885 Weismann suggested that hereditary characteristics were transmitted by what he called germ plasm—as distinguished from the somatoplasm (body cells) —which linked the generations by a continuous stream of dividing germ cells. In stating definitely seven years later that the material of heredity was in the chromosomes, Weismann anticipated the chromosomal basis of inheritance.

The English explorer, anthropologist, and eugenicist Francis Galton made a number of important contributions to genetics in the 19th century, one of which was a study of the hereditary nature of ability, from which he developed the concept that judicious breeding could improve the human race (eugenics). Galton's most-significant work was the demonstration that each generation of ancestors makes a proportionate contribution to the total makeup of the individual. Thus, he suggested, if a tall man marries a short woman, each should contribute half of the total heritage, and the resultant offspring should be intermediate between the two parents.

Mendelian Laws of Heredity

The fame of Gregor Mendel, the father of genetics, rests on experiments he did with garden peas, which possess sharply contrasting characteristics—for example, tall versus short; round seed versus wrinkled seed. When Mendel fertilized short plants with pollen from tall plants, he found the offspring (first filial generation) to be uniformly tall. But if he allowed the plants of that generation to self-pollinate (fertilize themselves), their offspring (the second filial generation) exhibited the characters of the grandparents in a rather consistent ratio of three tall to one short. Furthermore, if allowed to self-pollinate, the short plants always bred true—they never produced anything but short plants. From those results Mendel developed the concept of dominance, based on the supposition that each plant carried two trait units, one of which dominated the other. Nothing was known at that time about chromosomes or meiosis, yet Mendel deduced from his results that the trait units, later called genes, could be a kind of physical particle that was transmitted from one generation to another through the reproductive mechanism.

Mendel's most-important concept was the idea that the paired genes present in the parent separate or segregate during the formation of the gametes. Moreover, in later experiments in which he studied the inheritance of two pairs of traits, Mendel showed that one pair of genes is independent of another. Thus, the principles of segregation and of independent assortment were established.

Mendel's findings were ignored for 35 years, probably for two reasons. Because the distinguished Swiss botanist Karl Wilhelm von Nägeli failed to recognize the significance of the work after Mendel sent him the results, he did nothing to encourage Mendel. Nägeli's great prestige and the lack of his endorsement indirectly weighed against widespread recognition of Mendel's work. Moreover, when the work was published, little was known about the cell, and the processes of mitosis and meiosis were completely unknown. Mendel's work was finally rediscovered in 1900, when three botanists independently recognized the worth of his studies from their own research and cited his publication in their work.

Elucidation of the Hereditary Mechanism

By 1901 it was understood how the hereditary units postulated by Mendel are distributed; it was also known that the somatic (body) cells have a double, or diploid, complement of chromosomes, while the reproductive cells have a single, or haploid, chromosome number. The experimental demonstration of the chromosomal basis for heredity had been firmly established by the German cytologist Theodor Boveri soon after the turn of the century and subsequently confirmed by others. To account for the large number of observed hereditary characters, Boveri suggested that each chromosome in a pair can exchange the hereditary factors it carries with those of the other chromosome. At first, the American geneticist Thomas Hunt Morgan dismissed that concept. Later, however, when he found that it agreed with his own laboratory findings, Morgan and his collaborators assigned the hereditary units (genes) specific positions, or loci, within the chromosomes. With the genes established as the carriers of hereditary traits, the English biologist William Bateson coined the term *genetics* for the experimental study of heredity and evolution.

Biology in the 20th and 21st Centuries

Just as the 19th century can be considered the age of cellular biology, the 20th and 21st centuries were characterized primarily by developments in molecular biology.

Important Conceptual and Technological Developments

By utilizing modern methods of investigation, such as X-ray diffraction and electron microscopy, to explore levels of cellular organization beyond that visible with a light microscope—the ultrastructure of the cell—new concepts of cellular function were produced. As a result, the study of the molecular organization of the cell had tremendous impact on biology during the 20th and 21st centuries. It also led directly to the convergence of many different scientific disciplines in order to acquire a better understanding of life processes.

Technologies such as DNA sequencing and the polymerase chain reaction also were developed, allowing biologists to peer into the genetic blueprints that give rise to organisms. First-generation sequencing technologies emerged in the 1970s and were followed several decades later by so-called next-generation sequencing technologies, which were superior in speed and cost-efficiency. Next-generation sequencing provided researchers with massive amounts of genetic data, typically gigabases in size (1 gigabase = 1,000,000,000 base pairs of DNA). Bioinformatics, which linked biological data with tools and techniques for data analysis, storage, and distribution, became an increasingly important part of biological studies, particularly those involving very large sets of genetic data.

In the 1970s the development of recombinant DNA technology opened the way to genetic engineering, which enabled researchers to recombine nucleic acids and thereby modify organisms' genetic codes, giving the organisms new abilities or eliminating undesirable traits. Those developments were followed by advances in cloning technologies, which led to the generation in 1996 of Dolly the sheep, the first clone of an adult mammal. Together, recombinant DNA technology and reproductive cloning (the method used to produce a living animal clone) facilitated great progress in the development of genetically modified organisms (GMOs). Such organisms became crucial components of biomedical research, where genetically modified (GM) mice and other animals were developed to model certain human diseases, thereby facilitating the investigation of new therapies and the factors that cause disease. Recombinant DNA technology played a crucial role in the generation of GM crops, including pest-resistant forms of cotton and herbicide-resistant forms of maize (corn) and soybeans.

In the 1990s and early 2000s, researchers worldwide increasingly came together in consortiums and other collaborative groups to accomplish major feats in biology. The first major success of those efforts was the sequencing of the human genome, which was accomplished through the Human Genome Project (HGP). The HGP began in 1990, supported by the U.S. Department of Energy and the National Institutes of Health (NIH). NIH researchers later joined forces with Celera Genomics, a private-sector enterprise, and the project was completed in 2003. Other collaborative projects soon followed, including the International HapMap Project, an outgrowth of the HGP, and the 1000 Genomes Project, which built on data from the HapMap effort.

The 20th and 21st centuries also saw major advances in areas of biology dealing with ecosystems, the environment, and conservation. In the 20th century, scientists realized that humans are as dependent upon Earth's natural resources as are other animals. However, humans were contributing to the progressive destruction of the environment, in part because of an increase in population pressure and certain technological advances. Lifesaving advances in medicine, for example, had allowed people to live longer and resulted in a dramatic drop in death rates (primarily in developed countries), contributing to an explosive increase in the human population. Chemical contaminants introduced into the environment by manufacturing processes, pesticides, automobile emissions, and other means seriously endangered all forms of life. Hence, biologists began to pay much greater attention to the relationships of living things to each other as well as to their biotic and abiotic environments.

The growing significance of climate change and its impact on ecosystems fueled advances in ecology, as well as the development of fields such as conservation biology and conservation genetics. As in almost every other area of biology, molecular biology came to fulfill an important role in those fields, with techniques such as whole genome sequencing being used to gather information on the genetic diversity of populations of endangered species and techniques such as cloning and genome editing raising the possibility of someday resurrecting extinct species (a process known as de-extinction). Information on the DNA sequences of a wide range of species also aided progress in scientists' understanding of evolution and systematics (the study of evolutionary relationships and the diversification of life).

By 'life,' we mean a thing that can nourish itself and grow and decay.

~ ARISTOTLE, 384 BC TO 322 BC

Scientist, Philosopher

Intradisciplinary and Interdisciplinary Work

By the 21st century, there were many important categories in the biological sciences and hence numerous specialties within fields. Botany, zoology, and microbiology dealt with types of organisms and their relationships with each other. Such disciplines had long been subdivided into more-specialized categories—for example, ichthyology, the study of fishes, and algology, the study of algae. Disciplines such as embryology and physiology, which dealt with the development and function of an organism, were divided further according to the kind of organism studied—for example, invertebrate embryology and mammalian physiology. Many developments in physiology and embryology had resulted from studies in cell biology, biophysics, and biochemistry. Likewise, research in cell physiology and cytochemistry, along with ultrastructural studies, helped scientists correlate cell structure with function. Ecology, which focused on relationships between organisms and their environment, included both the physical features of the environment and other organisms that may compete for food and shelter. Emphasis on different environments and certain features of organisms resulted in the subdivision of the field into a range of specialties, such as freshwater ecology, marine ecology, and population ecology.

Many areas of study in the biological sciences cross the boundaries that traditionally separated the various branches of the sciences. In biophysics, for example, researchers apply the principles and methods of physics to investigate and find solutions to problems in biology. Evolutionary biologists and paleontologists are familiar with the principles of geology and may even work closely with geologists while attempting to determine the age of biological remains. Likewise, anthropologists and archaeologists apply knowledge of human culture and society to biological findings in order to more fully understand humankind. Astrobiology arose through the activities of the scientists and engineers concerned with the exploration of space. As a result, the field of biology has received contributions from and made contributions to many other disciplines, in the humanities as well as in the sciences.

Through the 20th and 21st centuries, as biology became increasingly interconnected with other areas of science, it also came to encompass a number of disciplines itself. In some of those disciplines, multiple levels of organization were recognized—for example, population biology (the study of populations of living things) and organismic biology (the study of the whole organism) and cell biology and molecular biology. In the latter part of the 20th century, molecular biology spawned still more disciplines, and the advent of genomics led to the emergence of sophisticated subdisciplines, such as developmental genomics and functional genomics. Genetics continued to expand, giving rise to new areas such as conservation genetics. Despite their diverse scope, however, in the 21st century many areas of the biological sciences continued to draw on common unifying principles and ideas, particularly those that were central to taxonomy, genetics, and evolution.

Changing Social and Scientific Values

In the 20th and 21st centuries, biologists' role in society as well as their moral and ethical responsibility in the discovery and development of new ideas led to a reassessment of individual social and scientific value systems. Scientists cannot afford to ignore the consequences of their discoveries; they are as concerned with the possible misuses of their findings as they are with the basic research in which they are involved. In the 20th century, the emerging social and political role of the biologist and all other scientists required a weighing of values that could not be done with the accuracy or objectivity of a laboratory balance. As members of society, it became necessary for biologists to redefine their social obligations and functions, particularly in the realm of making judgments about ethical problems, such as human control of the environment or the manipulation of genes to direct further evolutionary development.

Coping with the Problems of the Future

Of particular consequence in the biological sciences was the development of genetic engineering. In cases of genetic deficiencies and disease, genetic engineering opened up the possibility of correcting gene defects to restore physiological function, potentially improving patients' quality of life. Gene therapy, in which a normal gene would be introduced into an individual's genome in order to repair a disease-causing mutation, was one means by which researchers could potentially achieve that goal. However, the possibilities for misuse of genetic engineering were vast. There was significant concern, for example, about genetically modified organisms, particularly modified crops, and their impacts on human and environmental health. The emergence of cloning technologies, including somatic cell nuclear transfer, also raised concerns. The Declaration on Human Cloning passed in 2005 by the United Nations called upon member states to prohibit the cloning of humans, though it left open the pursuit of therapeutic cloning. Similarly, in 2015 researchers who had developed technologies for gene editing, which enabled scientists to customize an organism's genetic makeup by altering specific bases in its DNA sequence, called for a moratorium on the application of the technologies in humans. The impacts of gene editing on human genetics were unknown, and there were no regulations in place to guide its use. The debate over gene editing renewed earlier discussions about the ethical and social impacts of genetic engineering in humans, especially its potential to be used to alter traits such as intelligence and appearance.

Other challenges confronting biologists included the search for ways to curb environmental pollution without interfering with efforts to improve the quality of life for humankind. Contributing to the problem of pollution was the problem of surplus human population. A rise in global human population had placed greater demands on the land, especially in the area of food production, and had necessitated increases in the operations of modern industry, the waste products of which contributed to the pollution of air, water, and soil. To find solutions to global warming, pollution, and other environmental problems, biologists worked with social scientists and other members of society in order to determine the requirements necessary for maintaining a healthy and productive planet. For although many of humankind's present and future problems may seem to be essentially social, political, or economic in nature, they have biological ramifications that could affect the very existence of life itself.

On to the Nitty-Gritty

In this text, I will cover the topics covered in IB Standard Level and IB Higher Level and the number of hours dedicated to each topic along with what the IB expects you to understand in each topic.

IB Biology SL and HL Core

Both IB Biology SL and HL consist of the same core requirements that consist of the same number of hours. Both classes will cover the same 6 topics in the order listed below with the same subtopics listed below:

Topic #1: Cell Biology - 15 Hours for Both SL and HL

Subtopic	Subtopic Number	IB Points to Understand

Introduction to cells	1.1	• "According to the cell theory, living organisms are composed of cells." • "Organisms consisting of only one cell carry out all functions of life in that cell." • "Surface area to volume ratio is important in the limitation of cell size." • "Multicellular organisms have properties that emerge from the interaction of their cellular components." • "Specialized tissues can develop by cell differentiation in multicellular organisms." • "Differentiation involves the expression of some genes and not others in a cell's genome." • "The capacity of stem cells to divide and differentiate along different pathways is necessary in embryonic development and also makes stem cells suitable for therapeutic uses."
Ultrastructure of cells	1.2	• "Prokaryotes have a simple cell structure without compartmentalization." • "Eukaryotes have a compartmentalized cell structure." • "Electron microscopes have a much higher resolution than light microscopes."
Membrane structure	1.3	• "Phospholipids form bilayers in water due to the amphipathic properties of phospholipid molecules." • "Membrane proteins are diverse in terms of structure, position in the membrane and function." • "Cholesterol is a component of animal cell membranes."

Subtopic			
Membrane transport	1.4	•	"Particles move across membranes by simple diffusion, facilitated diffusion, osmosis and active transport."
		•	"The fluidity of membranes allows materials to be taken into cells by endocytosis or released by exocytosis. Vesicles move materials within cells."
The origin of cells	1.5	•	"Cells can only be formed by division of pre-existing cells."
		•	"The first cells must have arisen from non-living material."
		•	"The origin of eukaryotic cells can be explained by the endosymbiotic theory."
Cell division	1.6	•	"Mitosis is division of the nucleus into two genetically identical daughter nuclei."
		•	"Chromosomes condense by supercoiling during mitosis. -Cytokinesis occurs after mitosis and is different in plant and animal cells."
		•	"Interphase is a very active phase of the cell cycle with many processes occurring in the nucleus and cytoplasm."
		•	"Cyclins are involved in the control of the cell cycle."
		•	"Mutagens, oncogenes and metastasis are involved in the development of primary and secondary tumours."

Topic #2: Molecular Biology - 21 Hours for Both SL and HL

Subtopic	Subtopic Number	IB Points to Understand

Molecules to metabolism	2.1	•	"Molecular biology explains living processes in terms of the chemical substances involved."
		•	"Carbon atoms can form four covalent bonds allowing a diversity of stable compounds to exist."
		•	"Life is based on carbon compounds including carbohydrates, lipids, proteins and nucleic acids."
		•	"Metabolism is the web of all the enzyme-catalyzed reactions in a cell or organism."
		•	"Anabolism is the synthesis of complex molecules from simpler molecules including the formation of macromolecules from monomers by condensation reactions."
		•	"Catabolism is the breakdown of complex molecules into simpler molecules including the hydrolysis of macromolecules into monomers."
Water	2.2	•	"Water molecules are polar and hydrogen bonds form between them."
		•	"Hydrogen bonding and dipolarity explain the cohesive, adhesive, thermal and solvent properties of water."
		•	"Substances can be hydrophilic or hydrophobic."
Carbohydrates and lipids	2.3	•	"Monosaccharide monomers are linked together by condensation reactions to form disaccharides and polysaccharide polymers."
		•	"Fatty acids can be saturated, monounsaturated or polyunsaturated."
		•	"Unsaturated fatty acids can be cis or trans isomers."
		•	"Triglycerides are formed by condensation from three fatty acids and one glycerol."

Proteins	2.4	• "Amino acids are linked together by condensation to form polypeptides." • "There are 20 different amino acids in polypeptides synthesized on ribosomes." • "Amino acids can be linked together in any sequence giving a huge range of possible polypeptides." • "The amino acid sequence of polypeptides is coded for by genes." • "A protein may consist of a single polypeptide or more than one polypeptide linked together." • "The amino acid sequence determines the three-dimensional conformation of a protein." • "Living organisms synthesize many different proteins with a wide range of functions." • "Every individual has a unique proteome."
Enzymes	2.5	• "Enzymes have an active site to which specific substrates bind." • "Enzyme catalysis involves molecular motion and the collision of substrates with the active site." • "Temperature, pH and substrate concentration affect the rate of activity of enzymes." • "Enzymes can be denatured." • "Immobilized enzymes are widely used in industry."

Structure of DNA and RNA	2.6	• "The nucleic acids DNA and RNA are polymers of nucleotides." • "DNA differs from RNA in the number of strands present, the base composition and the type of pentose." • "DNA is a double helix made of two antiparallel strands of nucleotides linked by hydrogen bonding between complementary base pairs."
DNA replication, transcription and translation	2.7	• "The replication of DNA is semi-conservative and depends on complementary base pairing." • "Helicase unwinds the double helix and separates the two strands by breaking hydrogen bonds." • "DNA polymerase links nucleotides together to form a new strand, using the pre-existing strand as a template." • "Transcription is the synthesis of mRNA copied from the DNA base sequences by RNA polymerase." • "Translation is the synthesis of polypeptides on ribosomes." • "The amino acid sequence of polypeptides is determined by mRNA according to the genetic code." • "Codons of three bases on mRNA correspond to one amino acid in a polypeptide." • "Translation depends on complementary base pairing between codons on mRNA and anticodons on tRNA."

Cell respiration	2.8	• "Cell respiration is the controlled release of energy from organic compounds to produce ATP." • "ATP from cell respiration is immediately available as a source of energy in the cell." • "Anaerobic cell respiration gives a small yield of ATP from glucose." • "Aerobic cell respiration requires oxygen and gives a large yield of ATP from glucose."
Photosynthesis	2.9	• "Photosynthesis is the production of carbon compounds in cells using light energy." • "Visible light has a range of wavelengths with violet the shortest wavelength and red the longest." • "Chlorophyll absorbs red and blue light most effectively and reflects green light more than other colors." • "Oxygen is produced in photosynthesis from the photolysis of water." • "Energy is needed to produce carbohydrates and other carbon compounds from carbon dioxide." • "Temperature, light intensity and carbon dioxide concentration are possible limiting factors on the rate of photosynthesis."

Everything that human beings or living animals do is done by protein molecules. And therefore the kind of proteins that one has and therefore the ability one has is determined by the genes that one has.

~ *HAR GOBIND KHORANA, 1922 – 2011*

Molecular Biologist

Topic #3: Genetics - 15 Hours for Both SL and HL

Subtopic	Subtopic Number	IB Points to Understand
Genes	3.1	• "A gene is a heritable factor that consists of a length of DNA and influences a specific characteristic. • A gene occupies a specific position on a chromosome. • The various specific forms of a gene are alleles. • Alleles differ from each other by one or only a few bases. • New alleles are formed by mutation. • The genome is the whole of the genetic information of an organism. • The entire base sequence of human genes was sequenced in the Human Genome Project."

Chromosomes	3.2	• "Prokaryotes have one chromosome consisting of a circular DNA molecule. • Some prokaryotes also have plasmids but eukaryotes do not. -Eukaryote chromosomes are linear DNA molecules associated with histone proteins. • In a eukaryote species there are different chromosomes that carry different genes. • Homologous chromosomes carry the same sequence of genes but not necessarily the same alleles of those genes. -Diploid nuclei have pairs of homologous chromosomes. -Haploid nuclei have one chromosome of each pair. • The number of chromosomes is a characteristic feature of members of a species. • A karyogram shows the chromosomes of an organism in homologous pairs of decreasing length. • Sex is determined by sex chromosomes and autosomes are chromosomes that do not determine sex."

Meiosis	3.3	<ul><li>"One diploid nucleus divides by meiosis to produce four haploid nuclei.</li><li>The halving of the chromosome number allows a sexual life cycle with fusion of gametes.</li><li>DNA is replicated before meiosis so that all chromosomes consist of two sister chromatids.</li><li>The early stages of meiosis involve pairing of homologous chromosomes and crossing over followed by condensation.</li><li>Orientation of pairs of homologous chromosomes prior to separation is random.</li><li>Separation of pairs of homologous chromosomes in the first division of meiosis halves the chromosome number.</li><li>Crossing over and random orientation promotes genetic variation.</li><li>Fusion of gametes from different parents promotes genetic variation."</li></ul>

Inheritance	3.4	• "Mendel discovered the principles of inheritance with experiments in which large numbers of pea plants were crossed." • "Gametes are haploid so contain only one allele of each gene." • "The two alleles of each gene separate into different haploid daughter nuclei during meiosis." • "Fusion of gametes results in diploid zygotes with two alleles of each gene that may be the same allele or different alleles." • "Dominant alleles mask the effects of recessive alleles but co-dominant alleles have joint effects." • "Many genetic diseases in humans are due to recessive alleles of autosomal genes, although some genetic diseases are due to dominant or co-dominant alleles." • "Some genetic diseases are sex-linked. The pattern of inheritance is different with sex-linked genes due to their location on sex chromosomes." • "Many genetic diseases have been identified in humans but most are very rare." • "Radiation and mutagenic chemicals increase the mutation rate and can cause genetic diseases and cancer."

Subtopic	Subtopic Number	IB Points to Understand
Genetic modification and biotechnology	3.5	<ul><li>"Gel electrophoresis is used to separate proteins or fragments of DNA according to size."</li><li>"PCR can be used to amplify small amounts of DNA."</li><li>"DNA profiling involves comparison of DNA."</li><li>"Genetic modification is carried out by gene transfer between species."</li><li>"Clones are groups of genetically identical organisms, derived from a single original parent cell."</li><li>"Many plant species and some animal species have natural methods of cloning."</li><li>"Animals can be cloned at the embryo stage by breaking up the embryo into more than one group of cells."</li><li>"Methods have been developed for cloning adult animals using differentiated cells."</li></ul>

Topic #4: Ecology - 12 Hours for Both SL and HL

Subtopic	Subtopic Number	IB Points to Understand

Species, communities and ecosystems	4.1	<ul><li>"Species are groups of organisms that can potentially interbreed to produce fertile offspring."</li><li>"Members of a species may be reproductively isolated in separate populations."</li><li>"Species have either an autotrophic or heterotrophic method of nutrition (a few species have both methods)."</li><li>"Consumers are heterotrophs that feed on living organisms by ingestion."</li><li>"Detritivores are heterotrophs that obtain organic nutrients from detritus by internal digestion."</li><li>"Saprotrophs are heterotrophs that obtain organic nutrients from dead organisms by external digestion."</li><li>"A community is formed by populations of different species living together and interacting with each other."</li><li>"A community forms an ecosystem by its interactions with the abiotic environment."</li><li>"Autotrophs obtain inorganic nutrients from the abiotic environment."</li><li>"The supply of inorganic nutrients is maintained by nutrient cycling."</li><li>"Ecosystems have the potential to be sustainable over long periods of time."</li></ul>

Energy flow	4.2	• "Most ecosystems rely on a supply of energy from sunlight." • "Light energy is converted to chemical energy in carbon compounds by photosynthesis." • "Chemical energy in carbon compounds flows through food chains by means of feeding." • "Energy released from carbon compounds by respiration is used in living organisms and converted to heat." • "Living organisms cannot convert heat to other forms of energy. -Heat is lost from ecosystems." • "Energy losses between trophic levels restrict the length of food chains and the biomass of higher trophic levels."

Carbon cycling	4.3	<ul><li>"Autotrophs convert carbon dioxide into carbohydrates and other carbon compounds."</li><li>"In aquatic ecosystems carbon is present as dissolved carbon dioxide and hydrogen carbonate ions."</li><li>"Carbon dioxide diffuses from the atmosphere or water into autotrophs."</li><li>"Carbon dioxide is produced by respiration and diffuses out of organisms into water or the atmosphere."</li><li>"Methane is produced from organic matter in anaerobic conditions by methanogenic archaeans and some diffuses into the atmosphere or accumulates in the ground."</li><li>"Methane is oxidized to carbon dioxide and water in the atmosphere."</li><li>"Peat forms when organic matter is not fully decomposed because of acidic and/or anaerobic conditions in waterlogged soils."</li><li>"Partially decomposed organic matter from past geological eras was converted either into coal or into oil and gas that accumulate in porous rocks."</li><li>"Carbon dioxide is produced by the combustion of biomass and fossilized organic matter."</li><li>"Animals such as reef-building corals and mollusca have hard parts that are composed of calcium carbonate and can become fossilized in limestone."</li></ul>

Climate change	4.4	<ul><li>"Carbon dioxide and water vapor are the most significant greenhouse gases."</li><li>"Other gases including methane and nitrogen oxides have less impact."</li><li>"The impact of a gas depends on its ability to absorb long wave radiation as well as on its concentration in the atmosphere."</li><li>"The warmed Earth emits longer wavelength radiation (heat)."</li><li>"Longer wave radiation is absorbed by greenhouse gases that retain the heat in the atmosphere."</li><li>"Global temperatures and climate patterns are influenced by concentrations of greenhouse gases."</li><li>"There is a correlation between rising atmospheric concentrations of carbon dioxide since the start of the industrial revolution 200 years ago and average global temperatures."</li><li>"Recent increases in atmospheric carbon dioxide are largely due to increases in the combustion of fossilized organic matter."</li></ul>

Instead of being the biological center of the Universe, I believe our planet is just an assembly station, but one with a major advantage over most other places. The constant presence of liquid water almost everywhere on the Earth is a huge advantage for life, especially for assembling life into complex forms by the process we call 'evolution.'

~ FRED HOYLE, 1915 TO 2001

Astrophysicist

Topic #5: Evolution and Biodiversity - 12 Hours for Both SL and HL

Subtopic	Subtopic Number	IB Points to Understand

Evidence for evolution	5.1	• "Evolution occurs when heritable characteristics of a species change." • "The fossil record provides evidence for evolution." • "Selective breeding of domesticated animals shows that artificial selection can cause evolution." • "Evolution of homologous structures by adaptive radiation explains similarities in structure when there are differences in function." • "Populations of a species can gradually diverge into separate species by evolution." • "Continuous variation across the geographical range of related populations matches the concept of gradual divergence."
Natural selection	5.2	• "Natural selection can only occur if there is variation among members of the same species." • "Mutation, meiosis and sexual reproduction cause variation between individuals in a species." • "Adaptations are characteristics that make an individual suited to its environment and way of life." • "Species tend to produce more offspring than the environment can support." • "Individuals that are better adapted tend to survive and produce more offspring while the less well adapted tend to die or produce fewer offspring." • "Individuals that reproduce pass on characteristics to their offspring." • "Natural selection increases the frequency of characteristics that make individuals better adapted and decreases the frequency of other characteristics leading to changes within the species."

Classification of biodiversity	5.3	<ul><li>"The binomial system of names for species is universal among biologists and has been agreed and developed at a series of congresses."</li><li>"When species are discovered they are given scientific names using the binomial system."</li><li>"Taxonomists classify species using a hierarchy of taxa."</li><li>"All organisms are classified into three domains."</li><li>"The principal taxa for classifying eukaryotes are kingdom, phylum, class, order, family, genus and species."</li><li>"In a natural classification, the genus and accompanying higher taxa consist of all the species that have evolved from one common ancestral species."</li><li>"Taxonomists sometimes reclassify groups of species when new evidence shows that a previous taxon contains species that have evolved from different ancestral species."</li><li>"Natural classifications help in identification of species and allow the prediction of characteristics shared by species within a group."</li></ul>

Cladistics	5.4	• "A clade is a group of organisms that have evolved from a common ancestor." • "Evidence for which species are part of a clade can be obtained from the base sequences of a gene or the corresponding amino acid sequence of a protein." • "Sequence differences accumulate gradually so there is a positive correlation between the number of differences between two species and the time since they diverged from a common ancestor." • "Traits can be analogous or homologous." • "Cladograms are tree diagrams that show the most probable sequence of divergence in clades." • "Evidence from cladistics has shown that classifications of some groups based on structure did not correspond with the evolutionary origins of a group or species."

Topic #6: Human Physiology - 20 Hours for Both SL and HL

Subtopic	Subtopic Number	IB Points to Understand

Digestion and absorption	6.1		
		•	"The contraction of circular and longitudinal muscle of the small intestine mixes the food with enzymes and moves it along the gut."
		•	"The pancreas secretes enzymes into the lumen of the small intestine."
		•	"Enzymes digest most macromolecules in food into monomers in the small intestine."
		•	"Villi increase the surface area of epithelium over which absorption is carried out."
		•	"Villi absorb monomers formed by digestion as well as mineral ions and vitamins."
		•	"Different methods of membrane transport are required to absorb different nutrients."

The blood system	6.2	• "Arteries convey blood at high pressure from the ventricles to the tissues of the body." • "Arteries have muscle cells and elastic fibers in their walls." • "The muscle and elastic fibers assist in maintaining blood pressure between pump cycles." • "Blood flows through tissues in capillaries. Capillaries have permeable walls that allow exchange of materials between cells in the tissue and the blood in the capillary." • "Veins collect blood at low pressure from the tissues of the body and return it to the atria of the heart." • "Valves in veins and the heart ensure circulation of blood by preventing backflow." • "There is a separate circulation for the lungs." • "The heart beat is initiated by a group of specialized muscle cells in the right atrium called the sinoatrial node." • "The sinoatrial node acts as a pacemaker." • "The sinoatrial node sends out an electrical signal that stimulates contraction as it is propagated through the walls of the atria and then the walls of the ventricles." • "The heart rate can be increased or decreased by impulses brought to the heart through two nerves from the medulla of the brain." • "Epinephrine increases the heart rate to prepare for vigorous physical activity."

Defense against infectious disease	6.3	• "The skin and mucous membranes form a primary defense against pathogens that cause infectious disease." • "Cuts in the skin are sealed by blood clotting." • "Clotting factors are released from platelets." • "The cascade results in the rapid conversion of fibrinogen to fibrin by thrombin." • "Ingestion of pathogens by phagocytic white blood cells gives non-specific immunity to diseases." • "Production of antibodies by lymphocytes in response to particular pathogens gives specific immunity." • "Antibiotics block processes that occur in prokaryotic cells but not in eukaryotic cells." • "Viruses lack a metabolism and cannot therefore be treated with antibiotics. Some strains of bacteria have evolved with genes that confer resistance to antibiotics and some strains of bacteria have multiple resistance."

Gas exchange	6.4	<ul><li>"Ventilation maintains concentration gradients of oxygen and carbon dioxide between air in alveoli and blood flowing in adjacent capillaries."</li><li>"Type I pneumocytes are extremely thin alveolar cells that are adapted to carry out gas exchange."</li><li>"Type II pneumocytes secrete a solution containing surfactant that creates a moist surface inside the alveoli to prevent the sides of the alveolus adhering to each other by reducing surface tension."</li><li>"Air is carried to the lungs in the trachea and bronchi and then to the alveoli in bronchioles."</li><li>"Muscle contractions cause the pressure changes inside the thorax that force air in and out of the lungs to ventilate them."</li><li>"Different muscles are required for inspiration and expiration because muscles only do work when they contract."</li></ul>

Neurons and synapses	6.5	<ul><li>"Neurons transmit electrical impulses."</li><li>"The myelination of nerve fibers allows for saltatory conduction."</li><li>"Neurons pump sodium and potassium ions across their membranes to generate a resting potential."</li><li>"An action potential consists of depolarization and repolarization of the neuron."</li><li>"Nerve impulses are action potentials propagated along the axons of neurons."</li><li>"Propagation of nerve impulses is the result of local currents that cause each successive part of the axon to reach the threshold potential."</li><li>"Synapses are junctions between neurons and between neurons and receptor or effector cells."</li><li>"When presynaptic neurons are depolarized they release a neurotransmitter into the synapse."</li><li>"A nerve impulse is only initiated if the threshold potential is reached."</li></ul>

Hormones, homeostasis and reproduction	6.6	<ul><li>"Insulin and glucagon are secreted by β and α cells of the pancreas respectively to control blood glucose concentration."</li><li>"Thyroxin is secreted by the thyroid gland to regulate the metabolic rate and help control body temperature."</li><li>"Leptin is secreted by cells in adipose tissue and acts on the hypothalamus of the brain to inhibit appetite."</li><li>"Melatonin is secreted by the pineal gland to control circadian rhythms."</li><li>"A gene on the Y chromosome causes embryonic gonads to develop as testes and secrete testosterone."</li><li>"Testosterone causes pre-natal development of male genitalia and both sperm production and development of male secondary sexual characteristics during puberty."</li><li>"Estrogen and progesterone cause pre-natal development of female reproductive organs and female secondary sexual characteristics during puberty."</li><li>"The menstrual cycle is controlled by negative and positive feedback mechanisms involving ovarian and pituitary hormones."</li></ul>

The finding of the double helix thus brought us not only joy but great relief. It was unbelievably interesting and immediately allowed us to make a serious proposal for the mechanism of gene duplication.

~JAMES WATSON, 1928 – PRESENT

Geneticist

Additional Higher Level Topics

These classes are only for Higher Level students - 60 hours total for HL only

Topic #7: Nucleic Acids - 9 Hours for HL Only

Subtopic	Subtopic Number	IB Points to Understand

DNA structure and replication	7.1	• "Nucleosomes help to supercoil the DNA." • "DNA structure suggested a mechanism for DNA replication." • "DNA polymerases can only add nucleotides to the 3' end of a primer." • "DNA replication is continuous on the leading strand and discontinuous on the lagging strand." • "DNA replication is carried out by a complex system of enzymes." • "Some regions of DNA do not code for proteins but have other important functions."
Transcription and gene expression	7.2	• "Transcription occurs in a 5' to 3' direction." • "Nucleosomes help to regulate transcription in eukaryotes." • "Eukaryotic cells modify mRNA after transcription." • "Splicing of mRNA increases the number of different proteins an organism can produce." • "Gene expression is regulated by proteins that bind to specific base sequences in DNA." • "The environment of a cell and of an organism has an impact on gene expression."

Translation	7.3	• "Initiation of translation involves assembly of the components that carry out the process." • "Synthesis of the polypeptide involves a repeated cycle of events." • "Disassembly of the components follows termination of translation." • "Free ribosomes synthesize proteins for use primarily within the cell." • "Bound ribosomes synthesize proteins primarily for secretion or for use in lysosomes." • "Translation can occur immediately after transcription in prokaryotes due to the absence of a nuclear membrane." • "The sequence and number of amino acids in the polypeptide is the primary structure." • "The secondary structure is the formation of alpha helices and beta pleated sheets stabilized by hydrogen bonding." • "The tertiary structure is the further folding of the polypeptide stabilized by interactions between R groups." • "The quaternary structure exists in proteins with more than one polypeptide chain."

Topic #8: Metabolism, Cell Respiration, and Photosynthesis - 14 Hours for HL Only

Subtopic	Subtopic Number	IB Points to Understand

Topic	Section	Understandings
Metabolism	8.1	• "Metabolic pathways consist of chains and cycles of enzyme-catalyzed reactions." • "Enzymes lower the activation energy of the chemical reactions that they catalyze." • "Enzyme inhibitors can be competitive or non-competitive." • "Metabolic pathways can be controlled by end-product inhibition."
Cell respiration	8.2	• "Cell respiration involves the oxidation and reduction of electron carriers." • "Phosphorylation of molecules makes them less stable." • "In glycolysis, glucose is converted to pyruvate in the cytoplasm." • "Glycolysis gives a small net gain of ATP without the use of oxygen." • "In aerobic cell respiration pyruvate is decarboxylated and oxidized, and converted into acetyl compound and attached to coenzyme A to form acetyl coenzyme A in the link reaction." • "In the Krebs cycle, the oxidation of acetyl groups is coupled to the reduction of hydrogen carriers, liberating carbon dioxide." • "Energy released by oxidation reactions is carried to the cristae of the mitochondria by reduced NAD and FAD." • "Transfer of electrons between carriers in the electron transport chain in the membrane of the cristae is coupled to proton pumping." • "In chemiosmosis protons diffuse through ATP synthase to generate ATP." • "Oxygen is needed to bind with the free protons to maintain the hydrogen gradient, resulting in the formation of water." • "The structure of the mitochondrion is adapted to the function it performs."

Photosynthesis	8.3	<ul><li>"Light-dependent reactions take place in the intermembrane space of the thylakoids."</li><li>"Light-independent reactions take place in the stroma."</li><li>"Reduced NADP and ATP are produced in the light-dependent reactions."</li><li>"Absorption of light by photosystems generates excited electrons."</li><li>"Photolysis of water generates electrons for use in the light-dependent reactions."</li><li>"Transfer of excited electrons occurs between carriers in thylakoid membranes."</li><li>"Excited electrons from Photosystem II are used to contribute to generate a proton gradient."</li><li>"ATP synthase in thylakoids generates ATP using the proton gradient."</li><li>"Excited electrons from Photosystem I are used to reduce NADP."</li><li>"In the light-independent reactions a carboxylase catalyses the carboxylation of ribulose bisphosphate."</li><li>"Glycerate 3-phosphate is reduced to triose phosphate using reduced NADP and ATP."</li><li>"Triose phosphate is used to regenerate RuBP and produce carbohydrates."</li><li>"Ribulose bisphosphate is reformed using ATP."</li><li>"The structure of the chloroplast is adapted to its function in photosynthesis."</li></ul>

Topic #9: Plant Biology - 13 Hours for HL Only

Subtopic	Subtopic Number	IB Points to Understand
Transport in the xylem of plants	9.1	- "Transpiration is the inevitable consequence of gas exchange in the leaf. -Plants transport water from the roots to the leaves to replace losses from transpiration. - The cohesive property of water and the structure of the xylem vessels allow transport under tension. - The adhesive property of water and evaporation generate tension forces in leaf cell walls. - Active uptake of mineral ions in the roots causes absorption of water by osmosis.
Transport in the phloem of plants	9.2	- "Plants transport organic compounds from sources to sinks. -Incompressibility of water allows transport along hydrostatic pressure gradients. -Active transport is used to load organic compounds into phloem sieve tubes at the source. - High concentrations of solutes in the phloem at the source lead to water uptake by osmosis. -Raised hydrostatic pressure causes the contents of the phloem to flow towards sinks."

Growth in plants	9.3	- "Undifferentiated cells in the meristems of plants allow indeterminate growth. -Mitosis and cell division in the shoot apex provide cells needed for extension of the stem and development of leaves. -Plant hormones control growth in the shoot apex. -Plant shoots respond to the environment by tropisms. -Auxin efflux pumps can set up concentration gradients of auxin in plant tissue. -Auxin influences cell growth rates by changing the pattern of gene expression."
Reproduction in plants	9.4	- "Flowering involves a change in gene expression in the shoot apex. - The switch to flowering is a response to the length of light and dark periods in many plants. - Success in plant reproduction depends on pollination, fertilization and seed dispersal. - Most flowering plants use mutualistic relationships with pollinators in sexual reproduction.

Topic #10: Genetics and Evolution - 8 Hours for HL Only

Subtopic	Subtopic Number	IB Points to Understand
Meiosis	10.1	• "Chromosomes replicate in interphase before meiosis." • "Crossing over is the exchange of DNA material between non-sister homologous chromatids." • "Crossing over produces new combinations of alleles on the chromosomes of the haploid cells." • "Chiasmata formation between non-sister chromatids can result in an exchange of alleles." • "Homologous chromosomes separate in meiosis I." • "Sister chromatids separate in meiosis II." • "Independent assortment of genes is due to the random orientation of pairs of homologous chromosomes in meiosis I."
Inheritance	10.2	• "Gene loci are said to be linked if on the same chromosome." • "Unlinked genes segregate independently as a result of meiosis." • "Variation can be discrete or continuous." • "The phenotypes of polygenic characteristics tend to show continuous variation." • "Chi-squared tests are used to determine whether the difference between an observed and expected frequency distribution is statistically significant."

Gene pools and speciation	10.3	<ul><li>"A gene pool consists of all the genes and their different alleles, present in an interbreeding population."</li><li>"Evolution requires that allele frequencies change with time in populations."</li><li>"Reproductive isolation of populations can be temporal, behavioral or geographic."</li><li>"Speciation due to divergence of isolated populations can be gradual."</li><li>"Speciation can occur abruptly."</li></ul>

Topic #11: Animal Physiology - 16 Hours for HL Only

Subtopic	Subtopic Number	IB Points to Understand

Antibody production and vaccination	11.1	<ul><li>"Every organism has unique molecules on the surface of its cells."</li><li>"Pathogens can be species-specific although others can cross species barriers."</li><li>"B lymphocytes are activated by T lymphocytes in mammals."</li><li>"Activated B cells multiply to form clones of plasma cells and memory cells."</li><li>"Plasma cells secrete antibodies."</li><li>"Antibodies aid the destruction of pathogens."</li><li>"White cells release histamine in response to allergens."</li><li>"Histamines cause allergic symptoms."</li><li>"Immunity depends upon the persistence of memory cells."</li><li>"Vaccines contain antigens that trigger immunity but do not cause the disease."</li><li>"Fusion of a tumor cell with an antibody-producing plasma cell creates a hybridoma cell."</li><li>"Monoclonal antibodies are produced by hybridoma cells."</li></ul>

Movement	11.2	<ul><li>"Bones and exoskeletons provide anchorage for muscles and act as levers."</li><li>"Synovial joints allow certain movements but not others."</li><li>"Movement of the body requires muscles to work in antagonistic pairs."</li><li>"Skeletal muscle fibers are multinucleate and contain specialized endoplasmic reticulum."</li><li>"Muscle fibers contain many myofibrils."</li><li>"Each myofibril is made up of contractile sarcomeres."</li><li>"The contraction of the skeletal muscle is achieved by the sliding of actin and myosin filaments."</li><li>"ATP hydrolysis and cross bridge formation are necessary for the filaments to slide."</li><li>"Calcium ions and the proteins tropomyosin and troponin control muscle contractions."</li></ul>

The kidney and osmoregulation	11.3	<ul><li>"Animals are either osmoregulators or osmoconformers."</li><li>"The Malpighian tubule system in insects and the kidney carry out osmoregulation and removal of nitrogenous wastes."</li><li>"The composition of blood in the renal artery is different from that in the renal vein."</li><li>"The ultrastructure of the glomerulus and Bowman's capsule facilitate ultrafiltration."</li><li>"The proximal convoluted tubule selectively reabsorbs useful substances by active transport."</li><li>"The loop of Henle maintains hypertonic conditions in the medulla."</li><li>"ADH controls reabsorption of water in the collecting duct."</li><li>"The length of the loop of Henle is positively correlated with the need for water conservation in animals."</li><li>"The type of nitrogenous waste in animals is correlated with evolutionary history and habitat."</li></ul>

Sexual reproduction	11.4	<ul><li>"Spermatogenesis and oogenesis both involve mitosis, cell growth, two divisions of meiosis and differentiation."</li><li>"Processes in spermatogenesis and oogenesis result in different numbers of gametes with different amounts of cytoplasm."</li><li>"Fertilization in animals can be internal or external."</li><li>"Fertilization involves mechanisms that prevent polyspermy."</li><li>"Implantation of the blastocyst in the endometrium is essential for the continuation of pregnancy."</li><li>"HCG stimulates the ovary to secrete progesterone during early pregnancy."</li><li>"The placenta facilitates the exchange of materials between the mother and fetus."</li><li>"Estrogen and progesterone are secreted by the placenta once it has formed."</li><li>"Birth is mediated by positive feedback involving estrogen and oxytocin."</li></ul>

The development of biology is going to destroy to some extent our traditional grounds for ethical belief and it is not easy to see what to put in their place.

~FRANCIS CRICK, 1916 – 2004

Molecular Biologist

Options

As a part of the IB Biology class you cover an additional subject of your choosing from the list below (typically you don't choose, but rather your teacher does so for you). Whichever option you or your teacher chooses you will cover 3 or 4 topics (15 hours total) for SL and an additional 2 or 3 topics (25 hours total) for HL.

Option A: Neurobiology and Behavior - 15 Hours for SL and HL

Subtopic	Subtopic Number	IB Points to Understand
Neural development	A.1	• "The neural tube of embryonic chordates is formed by infolding of ectoderm followed by elongation of the tube." • "Neurons are initially produced by differentiation in the neural tube." • "Immature neurons migrate to a final location." • "An axon grows from each immature neuron in response to chemical stimuli." • "Some axons extend beyond the neural tube to reach other parts of the body." • "A developing neuron forms multiple synapses." • "Synapses that are not used do not persist." • "Neural pruning involves the loss of unused neurons." • "The plasticity of the nervous system allows it to change with experience."

The human brain	A.2	• "The anterior part of the neural tube expands to form the brain." • "Different parts of the brain have specific roles." • "The autonomic nervous system controls involuntary processes in the body using centers located mainly in the brain stem." • "The cerebral cortex forms a larger proportion of the brain and is more highly developed in humans than other animals." • "The human cerebral cortex has become enlarged principally by an increase in total area with extensive folding to accommodate it within the cranium." • "The cerebral hemispheres are responsible for higher order functions." • "The left cerebral hemisphere receives sensory input from sensory receptors in the right side of the body and the right side of the visual field in both eyes and vice versa for the right hemisphere." • "The left cerebral hemisphere controls muscle contraction in the right side of the body and vice versa for the right hemisphere." • "Brain metabolism requires large energy inputs."

| Perception of stimuli | A.3 | • "Receptors detect changes in the environment."
• "Rods and cones are photoreceptors located in the retina."
• "Rods and cones differ in their sensitivities to light intensities and wavelengths."
• "Bipolar cells send the impulses from rods and cones to ganglion cells."
• "Ganglion cells send messages to the brain via the optic nerve."
• "The information from the right field of vision from both eyes is sent to the left part of the visual cortex and vice versa."
• "Structures in the middle ear transmit and amplify sound."
• "Sensory hairs of the cochlea detect sounds of specific wavelengths."
• "Impulses caused by sound perception are transmitted to the brain via the auditory nerve."
• "Hair cells in the semicircular canals detect movement of the head." |
|---|---|---|

Additional Higher Level Topics - Additional 10 Hours for HL

Innate and learned behavior	A.4	<ul><li>"Innate behavior is inherited from parents and so develops independently of the environment."</li><li>"Autonomic and involuntary responses are referred to as reflexes."</li><li>"Reflex arcs comprise the neurons that mediate reflexes."</li><li>"Reflex conditioning involves forming new associations."</li><li>"Learned behavior develops as a result of experience."</li><li>"Imprinting is learning occurring at a particular life stage and is independent of the consequences of behavior."</li><li>"Operant conditioning is a form of learning that consists of trial and error experiences."</li><li>"Learning is the acquisition of skill or knowledge."</li><li>"Memory is the process of encoding, storing and accessing information."</li></ul>

Neuropharmacology	A.5	• "Some neurotransmitters excite nerve impulses in postsynaptic neurons and others inhibit them." • "Nerve impulses are initiated or inhibited in post-synaptic neurons as a result of summation of all excitatory and inhibitory neurotransmitters received from presynaptic neurones." • "Many different slow-acting neurotransmitters modulate fast synaptic transmission in the brain." • "Memory and learning involve changes in neurones caused by slow-acting neurotransmitters." • "Psychoactive drugs affect the brain by either increasing or decreasing postsynaptic transmission." • "Anesthetics act by interfering with neural transmission between areas of sensory perception and the CNS." • "Stimulant drugs mimic the stimulation provided by the sympathetic nervous system." • "Addiction can be affected by genetic predisposition, social environment and dopamine secretion."
Ethology	A.6	• "Ethology is the study of animal behavior in natural conditions." • "Natural selection can change the frequency of observed animal behavior." • "Behavior that increases the chances of survival and reproduction will become more prevalent in a population." • "Learned behavior can spread through a population or be lost from it more rapidly than innate behavior."

Option B: Biotechnology and Bioinformatics - 15 Hours for SL and HL

Subtopic	Subtopic Number	IB Points to Understand
Microbiology: organisms in industry	B.1	- "Microorganisms are metabolically diverse. - Microorganisms are used in industry because they are small and have a fast growth rate. - Pathway engineering optimizes genetic and regulatory processes within microorganisms. - Pathway engineering is used industrially to produce metabolites of interest. - Fermenters allow large-scale production of metabolites by microorganisms. -Fermentation is carried out by batch or continuous culture. - Microorganisms in fermenters become limited by their own waste products. - Probes are used to monitor conditions within fermenters. -Conditions are maintained at optimal levels for the growth of the microorganisms being cultured."

Biotechnology in agriculture	B.2	•- "Transgenic organisms produce proteins that were not previously part of their species' proteome. •-Genetic modification can be used to overcome environmental resistance to increase crop yields. •-Genetically modified crop plants can be used to produce novel products. •-Bioinformatics plays a role in identifying target genes. •- The target gene is linked to other sequences that control its expression. •- An open reading frame is a significant length of DNA from a start codon to a stop codon. •- Marker genes are used to indicate successful uptake. •-Recombinant DNA must be inserted into the plant cell and taken up by its chromosome or chloroplast DNA. •- Recombinant DNA can be introduced into whole plants, leaf discs or protoplasts. •-Recombinant DNA can be introduced by direct physical and chemical methods or indirectly by vectors."

Environmental protection	B.3	- "Responses to pollution incidents can involve bioremediation combined with physical and chemical procedures. -Microorganisms are used in bioremediation. -Some pollutants are metabolized by microorganisms. - Cooperative aggregates of microorganisms can form biofilms. -Biofilms possess emergent properties. - Microorganisms growing in a biofilm are highly resistant to antimicrobial agents. -Microorganisms in biofilms cooperate through quorum sensing. - Bacteriophages are used in the disinfection of water systems."

Nearly 2.5 billion years of prokaryotic cells and nothing else – two-thirds of life's history in stasis at the lowest level of recorded complexity... Why did life remain at stage 1 for two-thirds of its history if complexity offers such benefits?

~STEPHEN JAY GOULD, 1941 TO 2002

Paleontologist

Additional Higher Level Topics - Additional 10 Hours for HL

Medicine	B.4	<ul><li>"Infection by a pathogen can be detected by the presence of its genetic material or by its antigens."</li><li>"Predisposition to a genetic disease can be detected through the presence of markers."</li><li>"DNA microarrays can be used to test for genetic predisposition or to diagnose the disease."</li><li>"Metabolites that indicate disease can be detected in blood and urine."</li><li>"Tracking experiments are used to gain information about the localization and interaction of a desired protein."</li><li>"Biopharming uses genetically modified animals and plants to produce proteins for therapeutic use."</li><li>"Viral vectors can be used in gene therapy."</li></ul>

Bioinformatics	B.5	• "Databases allow scientists easy access to information." • "The body of data stored in databases is increasing exponentially." • "BLAST searches can identify similar sequences in different organisms." • "Gene function can be studied using model organisms with similar sequences." • "Sequence alignment software allows comparison of sequences from different organisms." • "BLASTn allows nucleotide sequence alignment while BLASTp allows protein alignment." • "Databases can be searched to compare newly identified sequences with sequences of known function in other organisms." • "Multiple sequence alignment is used in the study of phylogenetics." • "EST is an expressed sequence tag that can be used to identify potential genes."

Option C: Ecology and Conservation - 15 Hours for SL and HL

Subtopic	Subtopic Number	IB Points to Understand

Species and communities	C.1	• "The distribution of species is affected by limiting factors." • "Community structure can be strongly affected by keystone species." • "Each species plays a unique role within a community because of the unique combination of its spatial habitat and interactions with other species." • "Interactions between species in a community can be classified according to their effect." • "Two species cannot survive indefinitely in the same habitat if their niches are identical."
Communities and ecosystems	C.2	• "Most species occupy different trophic levels in multiple food chains." • "A food web shows all the possible food chains in a community." • "The percentage of ingested energy converted to biomass is dependent on the respiration rate." • "The type of stable ecosystem that will emerge in an area is predictable based on climate." • "In closed ecosystems energy but not matter is exchanged with the surroundings." • "Disturbance influences the structure and rate of change within ecosystems."

Impacts of humans on ecosystems	C.3	• "Introduced alien species can escape into local ecosystems and become invasive." • "Competitive exclusion and the absence of predators can lead to reduction in the numbers of endemic species when alien species become invasive." • "Pollutants become concentrated in the tissues of organisms at higher trophic levels by biomagnification." • "Macroplastic and microplastic debris has accumulated in marine environments."
Conservation of biodiversity	C.4	• "An indicator species is an organism used to assess a specific environmental condition." • "Relative numbers of indicator species can be used to calculate the value of a biotic index." • "In situ conservation may require active management of nature reserves or national parks." • "Ex situ conservation is the preservation of species outside their natural habitats." • "Biogeographic factors affect species diversity." • "Richness and evenness are components of biodiversity."

Additional Higher Level Topics - Additional 10 Hours for HL

Population ecology	C.5	• "Sampling techniques are used to estimate population size." • "The exponential growth pattern occurs in an ideal, unlimited environment." • "Population growth slows as a population reaches the carrying capacity of the environment." • "The phases shown in the sigmoid curve can be explained by relative rates of natality, mortality, immigration and emigration." • "Limiting factors can be top down or bottom up."
Nitrogen and phosphorus cycles	C.6	• "Nitrogen-fixing bacteria convert atmospheric nitrogen to ammonia." • "Rhizobium associates with roots in a mutualistic relationship." • "In the absence of oxygen denitrifying bacteria reduce nitrate in the soil." • "Phosphorus can be added to the phosphorus cycle by application of fertilizer or removed by the harvesting of agricultural crops." • "The rate of turnover in the phosphorus cycle is much lower than the nitrogen cycle." • "Availability of phosphate may become limiting to agriculture in the future." • "Leaching of mineral nutrients from agricultural land into rivers causes eutrophication and leads to increased biochemical oxygen demand."

Option D: Human Physiology - 15 Hours for SL and HL

Subtopic	Subtopic Number	IB Points to Understand
Human nutrition	D.1	• "Essential nutrients cannot be synthesized by the body, therefore they have to be included in the diet." • "Dietary minerals are essential chemical elements." • "Vitamins are chemically diverse carbon compounds that cannot be synthesized by the body." • "Some fatty acids and some amino acids are essential." • "Lack of essential amino acids affects the production of proteins." • "Malnutrition may be caused by a deficiency, imbalance or excess of nutrients in the diet." • "Appetite is controlled by a centre in the hypothalamus." • "Overweight individuals are more likely to suffer hypertension and type II diabetes." • "Starvation can lead to breakdown of body tissue."

| Digestion | D.2 | - "Nervous and hormonal mechanisms control the secretion of digestive juices."
- "Exocrine glands secrete to the surface of the body or the lumen of the gut."
- "The volume and content of gastric secretions are controlled by nervous and hormonal mechanisms."
- "Acid conditions in the stomach favor some hydrolysis reactions and help to control pathogens in ingested food."
- "The structure of cells of the epithelium of the villi is adapted to the absorption of food."
- "The rate of transit of materials through the large intestine is positively correlated with their fibre content."
- "Materials not absorbed are egested." |
|---|---|---|

Functions of the liver	D.3	<ul><li>"The liver removes toxins from the blood and detoxifies them."</li><li>"Components of red blood cells are recycled by the liver."</li><li>"The breakdown of erythrocytes starts with phagocytosis of red blood cells by Kupffer cells."</li><li>"Iron is carried to the bone marrow to produce hemoglobin in new red blood cells."</li><li>"Surplus cholesterol is converted to bile salts."</li><li>"Endoplasmic reticulum and Golgi apparatus in hepatocytes produce plasma proteins."</li><li>"The liver intercepts blood from the gut to regulate nutrient levels."</li><li>"Some nutrients in excess can be stored in the liver."</li></ul>

The heart	D.4	• "Structure of cardiac muscle cells allows propagation of stimuli through the heart wall." • "Signals from the sinoatrial node that cause contraction cannot pass directly from atria to ventricles." • "There is a delay between the arrival and passing on of a stimulus at the atrioventricular node." • "This delay allows time for atrial systole before the atrioventricular valves close." • "Conducting fibers ensure coordinated contraction of the entire ventricle wall." • "Normal heart sounds are caused by the atrioventricular valves and semilunar valves closing causing changes in blood flow."

I think it would be quite wrong to suggest that my colleagues have rejected me or that I reject them. Quite the reverse. It's only a small, vociferous group – mainly biologists, I'm sorry to say – that go beyond ordinary scientific criticism and start becoming personal.

~JAMES LOVELOCK, 1919 – PRESENT

Gaia Theorist, Inventor, Scientific Polymath

Additional Higher Level Topics - Additional 10 Hours for HL

Hormones and metabolism	D.5	<ul><li>"Endocrine glands secrete hormones directly into the bloodstream."</li><li>"Steroid hormones bind to receptor proteins in the cytoplasm of the target cell to form a receptor–hormone complex."</li><li>"The receptor–hormone complex promotes the transcription of specific genes."</li><li>"Peptide hormones bind to receptors in the plasma membrane of the target cell."</li><li>"Binding of hormones to membrane receptors activates a cascade mediated by a second messenger inside the cell."</li><li>"The hypothalamus controls hormone secretion by the anterior and posterior lobes of the pituitary gland."</li><li>"Hormones secreted by the pituitary control growth, developmental changes, reproduction and homeostasis."</li></ul>

Transport of respiratory gases	D.6	<ul><li>"Oxygen dissociation curves show the affinity of hemoglobin for oxygen."</li><li>"Carbon dioxide is carried in solution and bound to hemoglobin in the blood."</li><li>"Carbon dioxide is transformed in red blood cells into hydrogencarbonate ions."</li><li>"The Bohr shift explains the increased release of oxygen by hemoglobin in respiring tissues."</li><li>"Chemoreceptors are sensitive to changes in blood pH."</li><li>"The rate of ventilation is controlled by the respiratory control centre in the medulla oblongata."</li><li>"During exercise the rate of ventilation changes in response to the amount of CO2 in the blood."</li><li>"Fetal hemoglobin is different from adult hemoglobin allowing the transfer of oxygen in the placenta onto the fetal hemoglobin."</li></ul>

Practical Scheme of Work

The student also needs to complete experiments and experimental reports as a part of any IB Science course. For SL, there is 40 hours of material. For HL, there is 60 hours of material. Here are the activities:

- Practical activities - 20 hours for SL and 40 hours for HL
    - Lab work in class counts towards these hours
- Individual investigation (internal assessment-IA) - 10 hours for SL and HL
    - A lab project along with a report that counts as 20% of your IB exam scores (written exam counts for the other 80%)
- Group 4 Project - 10 hours for SL and HL
    - Students are separated into groups and must conduct an experiment and write a report.

Self-Test

Know **these Definitions**

Define is a 'simple' objective 1 command term... but you must be precise in your answers. Definitions are also a great start to review.

- Quiz yourself on the definitions, check your answers.

- Pay attention to the markschemes – what is the importance of the underlined terms and why can't you get marks without them?

- 'Unpack' the definition into its component parts – what is the relevance of each and how does it lead to more in-depth explanation of the concept?

Here's a quiz for the **define** assessment statements.

Give the precise meaning of a word, phrase or physical quantity.

1. **Define** *diffusion*.

....
....
....
....
....

2. **Define** *osmosis*.

....
....
....
....

3. **Define** *enzyme*.

....

....

....

....

....

4. **Define** *active site*.

....

....

....

....

....

5. **Define** *denaturation*.

....

....

....

....

....

6. **Define** *cell respiration*.

....
....
....
....
....

7. **Define** *gene*.

....
....
....
....
....

8. **Define** *allele*.

....
....
....
....
....

9. **Define** *genome*.

....
....
....
....
....

10. **Define** *gene mutation*.

....
....
....
....
....

11. **Define** *homologous chromosomes*.

....
....
....
....
....

12. **Define** *genotype*.

....
....
....
....
....

13. **Define** *phenotype*.

....
....
....
....
....

14. **Define** *dominant allele*.

....
....
....
....
....

15. **Define** *recessive allele*.

....
....

....

....

....

16. Define *codominant alleles*.

....

....

....

....

....

17. Define *homozygous*.

....

....

....

....

....

18. Define *heterozygous*.

....

....

....

....

....

19. **Define** *locus*.

....

....

....

....

....

20. **Define** *carrier*.

....

....

....

....

....

21. **Define** *test cross*.

....

....

....

....

22. **Define** *sex linkage*.

23. **Define** *clone*.

24. **Define** *species*.

25. **Define** *habitat*.

....
....
....
....
....

26. **Define** *population*.

....
....
....
....
....

27. **Define** *community*.

....
....
....
....
....

28. **Define** *ecosystem*.

....
....
....
....
....

29. **Define** *ecology*.

....
....
....
....
....

30. **Define** *trophic level*.

....
....
....
....
....

31. **Define** *evolution*.

32. **Define** *pathogen*.

33. **Define** *resting potential (repolarization)*.

34. **Define** *action potential (depolarization)*.

This is an era of specialists, each of whom sees his own problem and is unaware of or intolerant of the larger frame into which it fits.

~ RACHEL CARSON, 1907 TO 1964

Marine Biologist

OPTION A & E

35. **Define** *nutrient*.

36. **Define** *stimulus*.

....
....
....
....
....

37. **Define** *response*.

....
....
....
....
....

38. **Define** *reflex*.

....
....
....
....
....

Why the dinosaurs died out is not known, but it is supposed to be because they had minute brains and devoted themselves to the growth of weapons of offense in the shape of numerous horns.

~ BERTRAND RUSSELL, 1872 – 1970

Mathematician and Philosopher

ANSWER KEY

1. **Define** *diffusion*.

Diffusion is the passive movement of particles from a region of high concentration to a region of low concentration.

2. **Define** *osmosis*.

Osmosis is the passive movement of water molecules, across a partially permeable membrane, from a region of lower solute concentration to a region of higher solute concentration.

3. **Define** *enzyme*.

Enzyme: globular proteins that act as catalysts for biochemical reactions.

4. **Define** *active site*.

Active site: specific region on the surface of an enzyme to which substrates bind and which catalyses biochemical reactions.

5. **Define** *denaturation*.

Denaturation is a structural change in a protein that results in the loss (usually permanent) of its biological properties. Refer only to heat and pH as agents.

6. **Define** *cell respiration*.

Cell respiration is the controlled release of energy from organic compounds in cells to form ATP.

7. **Define** *gene*.

Gene: a heritable factor that controls a specific characteristic.

8. **Define** *allele*.

Allele: one specific form of a gene, differing from other alleles by one or a few bases only and occupying the same gene locus as other alleles of the gene.

9. **Define** *genome*.

Genome: the whole of the genetic information of an organism.

10. **Define** *gene mutation*.

A change in the base-sequence of an allele.

11. **Define** *homologous chromosomes*.

Chromosome pairs, one from each parent, that are identical or very similar in gene composition. Can be identified by size, banding pattern and centromere position.

12. **Define** *genotype*.

Genotype: the alleles of an organism.

13. **Define** *phenotype*.

Phenotype: the characteristics of an organism.

14. **Define** *dominant allele*.

Dominant allele: an allele that has the same effect on the phenotype whether it is present in the homozygous or heterozygous state.

15. **Define** *recessive allele*.

Recessive allele: an allele that only has an effect on the phenotype when present in the homozygous state.

16. **Define** *codominant alleles*.

Codominant alleles: pairs of alleles that both affect the phenotype when present in a heterozygote.

17. **Define** *homozygous*.

Homozygous: having two identical alleles of a gene.

18. **Define** *heterozygous*.

Homozygous: having two different alleles of a gene.

19. **Define** *locus*.

Locus: the particular position on homologous chromosomes of a gene.

20. **Define** *carrier*.

Carrier: an individual that has one copy of a recessive allele that causes a genetic disease in individuals that are homozygous for this allele.

21. **Define** *test cross*.

Test cross: testing a suspected heterozygote by crossing it with a known homozygous recessive.

22. **Define** *sex linkage*.

An association between genes in sex chromosomes that makes some traits more prevalent in on gender than the other. These genes are located on the non-homologous region of the X-chromosome in humans.

23. **Define** *clone*.

Clone: a group of genetically identical organisms or a group of cells derived from a single parent cell.

24. **Define** *species*.

Species: a group of organisms that can interbreed and produce fertile offspring.

25. Define *habitat*.

Habitat: the environment in which a species normally lives or the location of a living organism.

26. Define *population*.

Population: a group of organisms of the same species who live in the same area at the same time.

27. Define *community*.

Community: a group of populations living and interacting with each other in an area.

28. Define *ecosystem*.

Ecosystem: a community and its abiotic environment.

29. Define *ecology*.

Ecology: the study of relationships between living organisms and between organisms and their environment.

30. Define *trophic level*.

A class of organisms that share the same position in a food chain (such as producers, primary consumers etc)

31. **Define** *evolution*.

Evolution is the cumulative change in the heritable characteristics of a population.

32. **Define** *pathogen*.

Pathogen: an organism or virus that causes a disease.

33. **Define** *resting potential*.

Resting potential: the **electrical potential** across the plasma membrane of a cell that is not conducting an impulse.

34. **Define** *action potential (depolarization and repolarization)*.

Action potential: the reversal and restoration of the **electrical potential** across the plasma membrane of a cell, as an electrical impulse passes along it (depolarization and repolarization).

Perhaps it is not amiss to remark that the biologist may not hope to solve the ultimate problems of life any more than the chemist and physicist may hope to penetrate the final mysteries of existence in the non-living world.

~EDMUND BEECHER WILSON, 1856 TO 1939

Geneticist, Biologist

OPTION A & E

35. **Define** *nutrient*.

Nutrient: a **chemical substance** found in **foods** that is **used in the human body**.

36. **Define** *stimulus*.

A stimulus is a **change in the environment** (internal or external) that is detected by a receptor and elicits a response.

37. **Define** *response*.

A response to a change in (internal or external) environment, detected by a receptor.

38. **Define** *reflex*.

A reflex is a **rapid, unconscious** response.

SCIENCE

HELPS YOU PROVE OTHERS ARE DUMB

[anonymous quote]

Sources

i - biology (n.p.d.). Exam Skills. Retrieved from https://i-biology.net/ibdpbio/exam-skills/

International Baccalaureate (n.p.d.). , Programmes, Biology. Retrieved from http://www.ibo.org/programmes/diploma-programme/curriculum/sciences/biology/

Rogers, Kara; Green, Edna R.; Joshi, Susan Heyner (2015, October 16). Biology. Encyclopedia Britannica. Retrieved from https://www.britannica.com/science/biology

Seigel, Dora (2015, July 1). SAT / ACT Prep Online Guides and Tips. The Complete IB Biology Syllabus: SL and HL. Retrieved from http://blog.prepscholar.com/the-complete-ib-biology-syllabus-standard-level-sl-and-higher-level-hl

Illustrations and Quotes

The images and illustrations in this book are unburdened with copyright and available and authorized for use under (a) the Creative Commons Attribution 2.0 Generic license through Wikimedia Commons, other public domain, or open content sources, (b) are in the public domain because they have exceeded the copyright term, (c) are freely available throughout the Internet and other media, unidentifiable, or uncopyrightable. The author of this book does not claim ownership of the images or illustrations. Thus, the images and illustrations are free and may be used by anyone for any purpose following the requirement of attribution, which is referenced below. The attribution herein does not in any way suggest that the originator of the images or illustrations endorses the author of this book, the contents of this book, or the author's use of the work.

Einstein, Albert (n.p.d.). Quote retrieved from http://www.goodreads.com/quotes/tag/science, image from http://www.goodreads.com/photo/author/9810.Albert_Einstein?page=1&photo=326686. Earliest published variant this writer has found attributing it to Einstein is "If we knew what it was we were doing, it wouldn't be called 'research,' would it?" from p. 272 of *Natural Capitalism* from 1999 (to see it, go to https://www.amazon.com/Natural-Capitalism-Paul-Hawken/dp/0316353167 and "search inside the book" for the word "Einstein").

Famous Scientists (2016). Brilliant Biology Quotes. Retrieved from http://www.famousscientists.org/brilliant-biology-quotes/

Front Cover (2016, September, 9). Human derived cells. SK8/18-2. Vimentin stained with Alexa 488 (green), nuclei stained with DAPI (blue). Imaged with ZEISS Axio Observer 7, Axiocam 506 mono and Colibri 7. Images donated as part of a GLAM collaboration with Carl Zeiss Microscopy. Retrieved from https://upload.wikimedia.org/wikipedia/commons/a/ad/SK8-18-2_human_derived_cells%2C_fluorescence_microscopy_%2829942101073%29.jpg

Science Helps You Prove... (n.p.d.) Funny Biology Quotes. Retrieved from http://quotesgram.com/img/funny-biology-quotes/5170889/

The End

Fine

La fin

El Fin

Das Ende

Конец

終わり

結束

النهاية

$$E^2 - |\vec{p}|^2 c^2 = m^2 c^4$$

AMAZING IB BOOKS AVAILABLE AT AMAZON AND OTHER FINE BOOK EMPORIUMS!

IB History Exam Study Guide: International Contemporary History 1848-2008 by Dr. Juan R. Céspedes, Ph.D. A comprehensive study guide for students preparing for the IB examination, extensively covering material on papers 1, 2, & 3. Also an excellent review source for any student of contemporary history (years 1848-2008). Examines the root causes of events leading to the history of the 20th century, as well as thought provoking material necessary for the critical thinking aspects found in IB examinations. Dr. Juan R. Céspedes' scholarly perspective is concise and covers the following topics: "Communism in Crisis: 1976-1991", "Causes, Practices and Effects of Wars", "Origins and Development of Authoritarian and Single Party States", "The Cold War", and "History of the Americas".

IB Paper 1-2: Collapse of the Soviet Empire - A Preparation for the International Baccalaureate History Examination!

An in-depth examination of the causes of the collapse of communism in the Soviet Union and Eastern Europe. An excellent review source for the IB history examination or for the historical researcher. Comprehensive and packed with IB-related thought provoking questions and material. Dr. Juan R. Céspedes' scholarly perspective engages the reader with details of the cultural, economic, social, and political life under communism. A sampling of the subject matter covered is "The Decline and Disintegration of the Soviet Union", "Achievements at Home and Abroad", "Outlook for Brezhnev", "Incursion into Afghanistan", "Enter Gorbachev", "Glasnost", "Perestroika", "Attempts at Democratization", "A Chronology of Key Events", "Possible Exam-Related Questions and Outline of Answers" and much more. Each chapter ends with a synopsis of the material covered for easy review!

The IB Extended Essay: An "A+" in 6 Easy Steps!

The Extended Essay is one of the core requirements of the IB Diploma Programme. Written on a topic chosen by the student, it culminates in a 4,000 word essay. This seems to be a daunting task for many students, but Dr. Céspedes breaks this seemingly difficult task into 6 simple steps. Dr. Céspedes has helped thousands of students with their writing, including doctoral candidates! Inside are tips for selecting a top-scoring topic, researching quickly and effectively, structuring your essay for maximum impact, and concluding impressively. Follow the step-by-step instructions in this guide and you will maximize your final score!

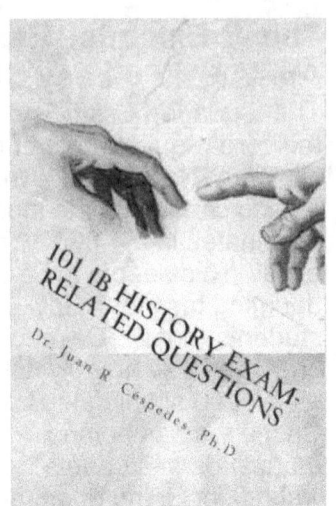

101 IB History Exam-related Questions: ...and their answers!

101 IB History Exam-related Questions! This book is a terrific resource for helping both students and teachers prepare for the demanding IB history examinations SL and HL. With this book you will learn from an experienced IB instructor and college professor, with 30 years in the educational field. You will improve your understanding of how questions should be answered in order to receive the highest possible scores. This book will help you build confidence by answering the questions provided, then checking your answers with those in the book. The book is great at covering the historical material for each question. Finally, it will help you understand the IB history syllabus and learn the proper way to approach your exams!

The APUSH Cruncher: A Guide for Passing the AP American History Exam with Ease!

As its name implies, The APUSH CRUNCHER is specifically designed for crunching the information you need to master the AP US History exam. From Pre-Columbian societies to 20th Century politics, this AP US History prep is designed for either short-term intense review or lengthier study. You have a choice! Develop your AP history skills in every test area. The CRUNCHER coaches you on the most important parts of the test, the weight given to each section, and proven strategies on how to effectively answer the questions for top scores. The author's goal is for you become a star-spangled winner in AP US history. This book is written by Dr. Juan R. Céspedes, a Ph.D. in education with over 30 years of experience!

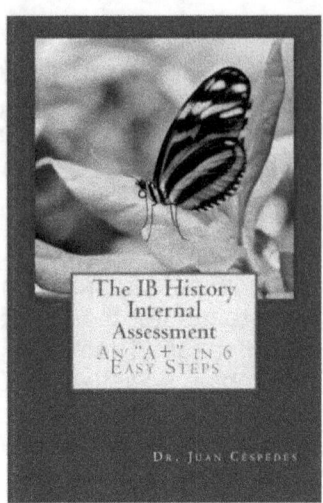

The IB History Internal Assessment: An "A+" in 6 Easy Steps!

The Internal Assessment is one of the core requirements of the IB Diploma Programme. Written on a topic chosen by the student, it culminates in a 2,000 word essay with specific requirements. These requirements often makes the Internal Assessment a difficult task for many students, but Dr. Céspedes breaks this into 6 simple steps. Dr. Céspedes has helped thousands of students with their writing, including doctoral candidates at the university level! Inside are tips for selecting a top-scoring topic, researching quickly and effectively, structuring your essay for maximum impact, and concluding impressively. Follow the step-by-step instructions in this guide and you will maximize your final score.

IB Paper 1: Communism - The Myopic Vision: The Causes of Totalitarianism, Authoritarianism, and Statism!

In this benchmark of political and historical analysis, Dr. Juan R. Céspedes provides a fascinating and comprehensive look at the "Myopic Vision" which has resulted in totalitarianism, authoritarianism, and statism. The Myopic Vision has inspired many governments and is responsible for the deaths of approximately 250 million human beings in the 20th century. What are the reasons behind this slaughter? The advocates of the Myopic Vision believe they are the holders of scientifically demonstrable truths concerning man, history, and social evolution. Thus, Dr. Céspedes coins a new phrase, as well as issues a warning for the future. Can the Myopic Vision be corrected or reversed? What does the future hold? This is a "must read" for all those which cherish and wish to preserve democracy and human dignity.

IB Paper 2: War Interminable: The Origins, Causes, Practices and Effects of International Conflict!

The Extended Essay is one of the A definitive and comprehensive examination on the origins, causes, practices and effects of warfare. An excellent review source for the IB history examination or for those interested in historical research. A sampling of the subject matter covered is "The Origins of War"; the "Anthropological Roots of War"; "The Nation-State and War"; the "School of Realism in International Relations"; "Zero-sum" gamesmanship; "National Interest and the Rule of Law"; "Waging War"; "Military Strategy & Planning"; war from the perspective of Clausewitz, Sun Tzu, Napoleon, Machiavelli, and other great thinkers; "The Causes of War"; "The Nature of War"; the role of women; "The Practice of War"; "Different types of 20th century warfare"; "The Effects and Results of War"; "Post-war economic problems"..."Possible IB Exam-Related Questions" and much more. Each chapter ends with a synopsis of the material covered for easy review.

Cambridge History Exam Prep: Maximize your score in 5 days!

Scientifically, the very best way to prepare for the Cambridge History exam! Studying for the Cambridge history exam can be a stressful time for all students– there is so much information to cover! So, knowing HOW to properly prepare for the exam is the key to avoiding stress and maximizing your score. This guide provides a clear and uniform way to focus and make the best use of your study time, while assessing your performance. Say "no" to cramming blindly. Focus specifically on questions that are likely to come up in the exam! Maximize your retention; this guide makes clear the sometimes missed associations between the curriculum covered in Cambridge and the areas of examination. Avoid the "all-nighter" which wastes time, impairs reasoning and memory. The summations provided in this guide are the most effective way to study.

Footsteps to World War III

A riveting, enlightening and often frightening look at the next major war involving the United States and the world. In his thought-provoking book, Dr. Céspedes draws from a multitude of sources to carefully analyze the likelihood, and the outcomes, of a major military encounter between the United States and its major potential adversaries. Asserting that we are at the doorstep of a new epoch, he constructs a lucid and highly comprehensible forecast of the challenges that the United States can expect from around the world. As a Europe continues to weaken militarily and chose neutralist policies, or policies contrary to those of the United States, what new alliances can the United States formulate? With this panorama, will the United States remain the dominant global superpower, as other nations challenge American preeminence? Will such a war ultimately end with a victory by the United States and its allies? Captivating and compelling from the first to the last page, "Footsteps to World War III" is a fascinating exploration of what the future may hold for the world at large.

The New Global Threat: Transnational Crime

Transnational organized crime is a growing cancer, and a threat to the security of North America . . . as well as to the entire world! What was once the dominion of countries and regions separated by geography is increasingly a concern to the Department of State and for the U.S. military. Transgressing the laws of multiple nations, these criminal activities have a negative impact on society, the global economy, and greatly impede the progress of developing nations and governments. Unfortunately, many of the nations affected lack the resources or the political will to provide an adequate level of security and countermeasures that matches the threat to its security and its borders. The criminal activities, such as drug, human, and gun trafficking, are often overshadowed by terrorism, even though all are crucial national security threats. This work reveals the dangers and horrors of modern transnational crime; a menace which challenges the modern hierarchies and methods of response, and helps to piece together the construction of pragmatic and workable policies to counter such a menace. This is a must read for anyone interested in national security issues!

Victory Over Terrorism: The Unthinkable Solution!

Extremists have made, and will continue into the future, varied attempts to attack the US and its allies with varying levels of organization and skill. We shake our heads with a certain degree of disbelief and say, "When will it end?" After the destruction of the World Trade Center and the accompanying damages to the Pentagon and the national psyche, government has enacted laws and strategies to prevent terrorist activities. However, as democratic societies which value individual rights, we are challenged to not provoke a response which undermines our Constitutional system of government. In a thorough investigative fashion, "Victory Over Terrorism" takes readers through the strange and often sordid methods that are routinely utilized by terrorists organizations, the prevailing thoughts and beliefs of Muslim communities throughout the world, and the anti-terrorist practices that intensified in recent years.

The TOK Essential Compendium!

An indispensable guide that provides an inside understanding of the TOK program from John R. Garden, an instructor with over 30 years of teaching experience! Centered around the "How We Know" and "Ways of Knowing" of 33 essential philosophers from Laozi to Albert Camus. Provides 350 TOK-oriented thinking questions, a blueprint for a top scoring essay, and blueprint for a top scoring presentation. A practical, easy to understand, and thorough examination of the Theory of Knowledge program.

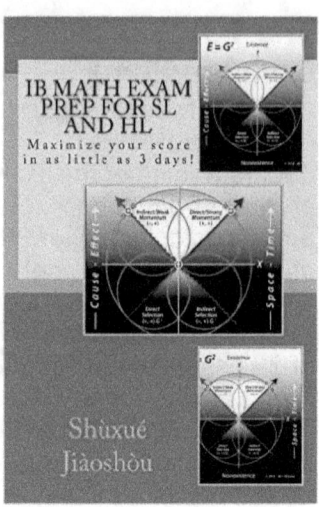

IB MATH EXAM PREP for SL and HL

How you prepare for the IB math exam is more important than how many hours you spend preparing! This study guide provides everything you need to master the challenging concepts from basic Algebra to Calculus, and will help you focus your studies on the most important math topics to maximize your score!

This comprehensive study guide contains many essential and unique features to help improve exam scores, including:

* Detailed explanations for solving each formula presented
* Methods and strategies to improve your math score
* Review of important Math Concepts

This study guide provides you with everything you need to improve your Math score—unquestionably. These test-taking techniques, methods, and strategies work. This study guide is the must-have preparation tool for every student looking to score higher on the IB Math Exams and get into their top-choice college!

Cuban Communism and its Impact on World Affairs: 1959 - 2015 (for IB the Americas - Paper 3)

Cuba is a relatively small country, but with the foreign policy of a major league international player. It has carried out such a policy since the start of Castro's 1959 revolution. Because of its close alliance with the Soviet Union, by the early to mid 1960's it had the external resources, internal conditions, and lack of significant US opposition to begin to visibly impact conditions throughout the world; from Latin America, to Africa and Asia. In this book, Dr. Céspedes sees a Cuban foreign policy that has been both aggressive ideologically, militaristic, yet highly pragmatic. The Castro brothers saturated the media with altruistic images of themselves in a campaign to justify and encourage, or directly spread, radical change. Using previously unexplored sources, Dr. Céspedes constructs a compelling and detailed exposé which focuses on Cuban political subversion throughout the world, either through direct military intervention, or disguised as development aid for health, economic, and civic programs. This is a must read for history students written by a prominent historian.

Cancer Free!

Your physician tells you the bad news: "You have cancer." The shock leaves you speechless, but you must go on. This book will give you hope. It is "must" reading for those that have cancer, or those with family members who have the disease. Motivated by a family history of cancer that caused the deaths of his grandfather, father, uncle, and eventually his wife, Dr. John Hunter engages in a personal quest to fully research and provide answers to dealing with this dreaded disease. Dr. Hunter explains which are the most advantageous options from the thousands of case-control studies, comprehensive reviews, and biologic parameters published on the topic, while informing the reader of the reasons for those selections. Dr. John Hunter sorts through the morass of disinformation that often accompanies the topic and sheds light on a complex disease, providing the public with interconnected facts and whole truths. Dr. Hunter accurately describes the current status of scientific and holistic thinking on many topics included in his book. He describes a multitude of strategies from quitting smoking to learning the warning signs of cancer.

Conversations with a Centurion: about Life, Death, and God

We struggle with work and overcoming the challenges of daily life, only to experience the same end...death. Rich or poor, educated or not, death is the great equalizer for all of humanity. One measure of adulthood is the realization that "bad" things happen to "good" people. Why am I here? Is there a God, what is the meaning of life, is there an afterlife? If there is a God...why does God let these things happen? These questions are answered in a meeting with a Roman Centurion that appears magically, mystically, inexplicably, in a museum in modern Rome. The Centurion turns out to be a Roman army officer who was assigned to Capernaum, a fishing village in Judea, where he met Jesus. Most people in our time, devout Christians and Jews included, understand little about the world of first-century Palestine, and think that the Jews were an undifferentiated and united people in their disgruntlement and opposition to the rule of Rome. This is a gross oversimplification of the historical, political, and cultural reality of the time. The Centurion helps the reader understand the life of Romans, Jews, their leaders, and others; as well as the geographical, cultural, and political divisions which existed at the time. The book examines many aspects of Jesus' world that are unknown to many. Dr. Céspedes also delves into mathematical and scientific explanations for the existence of God, near death experiences, the crucifixion, the resurrection of Jesus, the early Christian church, and the great personal sacrifices made by the apostles. This fascinating and carefully researched book concludes with suggestions for living a more meaningful life, and how to battle sadness and loneliness.

www.ingramcontent.com/pod-product-compliance
Lightning Source LLC
Chambersburg PA
CBHW070251190526
45169CB00001B/372